Hot Work Tool Steel

Joachim Schlegel · Till Schneiders

Hot Work Tool Steel

A Steel Portrait

Springer Vieweg

Joachim Schlegel
Hartmannsdorf, Sachsen, Germany

Till Schneiders
Herne, Germany

ISBN 978-3-658-43015-3   ISBN 978-3-658-43016-0 (eBook)
https://doi.org/10.1007/978-3-658-43016-0

This Springer Vieweg imprint is published by the registered company Springer Fachmedien Wiesbaden GmbH, part of Springer Nature.
The registered company address is: Abraham-Lincoln-Str. 46, 65189 Wiesbaden, Germany

Paper in this product is recyclable.

Preface

Steel is indispensable, recyclable, and has a very special significance: In our modern industrial society, steel is the basic material for all major industrial sectors, and even today's global megatrends, such as climate change, mobility, and healthcare, cannot be solved or managed without steel.

The 5000-year long history of iron and steel production is impressive. The world of steel has become astonishingly diverse and so complex that it is not easy to grasp in practice (Schlegel 2021). In the form of *essentials* and *brochures* for portraits of selected steels and steel groups, this world of steel is to be brought closer to the reader; compact, understandable, informative, structured with examples from practice, and suitable for reference.

This brochure describes the **hot work tool steels**, a group of alloyed steels suitable for tools with high heat resistance. These can withstand surface temperatures of more than 600 °C in use. For this purpose, they are optimally adapted to the most diverse requirements, especially for tools for hot forming and die casting. The chemical compositions, manufacturing and processing methods, as well as the properties and material data of hot work tool steels, are briefly and clearly presented.

We would like to thank Mr. Frieder Kumm M.A., Senior Editor of the Construction Engineering Department of Springer Vieweg Publishing, for his motivation, supervision, and support. We also thank Mr. Dr. Christian Schlegel and Mr. Dr. Peter Schlegel for their help with proofreading the manuscript in German and in English, respectively.

Hartmannsdorf, Germany   Dr.-Ing. Joachim Schlegel
Herne, Germany   Dr.-Ing. Till Schneiders

v

Table of Contents

Fundamentals 1

1.1 What is a Hot Work Tool Steel?

A hot work tool steel is an alloyed tool steel that, as the name suggests, is suitable for "hot working" of materials.

Hot work tool steels are used for tools for the non-cutting shaping of materials at surface temperatures of the tool above 200 °C (DIN EN ISO 4957). The workpiece temperatures can vary between 400 and 1200 °C. The main areas of application are die casting molds, extrusion dies, and forging tools. The hot work tool steel withstands complex mechanical, thermal, chemical, and tribological stresses. Due to these different loads, hot work tool steels are divided into three groups. The first group includes martensitic steels with low secondary hardness. The steels of the second group have a higher alloy content and pronounced secondary hardness. For long contact duration and simultaneously high workpiece temperature, such as during extrusion of heavy metals, the heat-resistant and scale-resistant austenitic steels of the third group are used.

1.2 On the History

Interesting aspects of the history of hot work tool steels are always related to the historical development of steels in general and to the manufacturing technology with its growing requirements. The knowledge gained about the effectiveness of alloying elements, especially regarding hot strength, plays a significant role; as well as the hardening process as an essential process in the production of tools made of hot work tool steels.

J. Schlegel and T. Schneiders, *Hot Work Tool Steel*, https://doi.org/10.1007/978-3-658-43016-0_1

Until about 1900, unalloyed carbon steels were used for tools. Alloyed tool steels with significantly better wear properties did not yet exist. These unalloyed steels were already hardened but lost their hardness at higher temperatures around 200 °C and wore out quite quickly. The essential hardening process had long been used by blacksmiths in China and Japan. Even before 900 AD hardened Japanese swords became famous, forged from soft and hard iron, i.e., iron with low and high carbon content. Iron was also known and used for weapons and utensils in Iraq, Cyprus, Egypt, Persia, Greece, among the Hittites and Etruscans, and many other peoples (Johannsen, 1953). The Iron Age in Europe began around 800 to 700 BC, a time when iron was still reduced from ore in earth pit furnaces (https://en.wikipedia.org/wiki/Iron_Age). Later, shaft furnaces, piece furnaces, raft furnaces, and finally blast furnaces were used. However, this took over 2000 years.

One of the many milestones, especially for tool steels, was the crucible steel process developed by *Benjamin Huntsman* (1704–1776) around 1740. A high-quality, very homogeneous steel could be produced, especially regarding carbon distribution, with more uniform properties (Spur, 1991).

Until the proof of the effect of carbon in steel around 1816, further, later so important metallic alloying elements for hot work tool steels had been discovered which would become increasingly important, such as cobalt (Co) in 1735, nickel (Ni) in 1751, tungsten (W) in 1783, molybdenum (Mo) in 1781, titanium (Ti) in 1791, and vanadium (V) in 1801. Finally, around 1854, *Robert Wilhelm Bunsen* (1811–1899) succeeded in producing pure chromium. From about 1850 onwards, research began into the effects of tungsten, chromium, and molybdenum (carbide formers) as well as other alloying elements in steel. For example, British metallurgist *Robert Mushet* (1811–1891) invented an improved tool steel with a 5% tungsten content, patented in 1861 (Ernst, 2009). And with a specially hardened, high-alloy chromium-tungsten steel, a groundbreaking success in metalworking was achieved. It was the invention of the "miracle turning steel" that *Frederick Winslow Taylor* (1856–1915) presented in 1900 at the World Exhibition in Paris as "**H**igh **S**peed **S**teel" (**HSS**) (Trent & Wright, 2000), (Ernst, 2009). It retains its hardness up to just below 600 °C and still functions as a tool steel even when glowing red.

Soon, the systematic development of tool steels began. The plant technology for steel production was also improved, and research into new steel production processes was intensified. For example, the first vanadium-containing steel came onto the market in England around 1903 (Bauer, 2000). In 1904, electric steel production began, followed by the era of stainless steels from 1912, the use of vacuum melting processes from 1928, and the electro-slag remelting of steel from

1930. From 1940 onwards, molybdenum was increasingly used as an alloying element in tool steel instead of tungsten. Since the 1960s, industrially produced powder metallurgical tool steels have been known, with hot work tool steels being produced in this way only since the 1980s (Bayer & Seilstorfer, 1984). And since that time, surface treatments and coatings have also been used to improve the wear resistance of forming, cutting, and die-casting tools.

The advances made in recent decades to further increase the performance of hot work tool steels can be divided into the development of new or the variation of existing alloys and the further development of production technologies. These include steel production in electric arc furnaces, improved secondary metallurgy in ladle furnaces, vacuum degassing plants, or AOD converters (decarburization with argon-oxygen mixture), as well as remelting processes (ESR—electro-slag remelting, VAR—vacuum arc remelting) to achieve very high purity of hot work tool steel. The reduction of residual oxygen and sulfur contents and the targeted influence of non-metallic inclusions in terms of quantitiy, size, and chemical composition e. g. through targeted calcium treatment in ladle metallurgy also ensure today's high standard of steel purity (Meyer et al., 1995), (Huemer, 2005). Finally, in addition to the chemical composition and metallurgical production and processing (forming), heat treatments are also important for achieving the desired steel properties (Liedtke, 2005), see Sect. 4.4: *Heat treatment*. Considering all these advanced technologies, the progress made with them in the last 60 years becomes apparent, for example, in the increase in toughness of hot work tool steels, accompanied by increasing purity through remelting (Ehrhardt, 2008). And the toughness (determined, for example, by means of impact bending tests) provides, in addition to ductility and hardness, an orientation for the service life of forming tools and die casting dies, as it helps to avoid heat or stress cracks and to delay the course of thermal fatigue. Figure 1.1 shows this trend since 1960 with references to the technologies used.

The great interest in hot work tool steels today and their importance can be seen from the international tool steel conferences held since 1987, at which these steels always formed an important focus (Schneiders, 2005). And for the future, hot work steel offers great potential and application possibilities in many industries (see Chap. 5: *Applications*), also demanded by innovative applications such as press hardening.

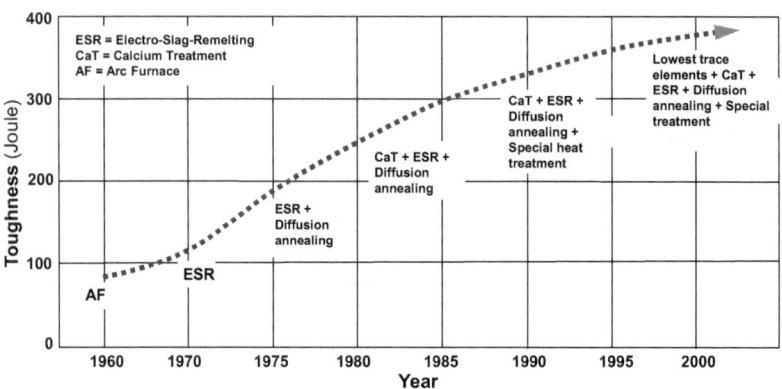

Fig. 1.1 Effects of manufacturing technologies on the toughness of hot work tool steel—trend since 1960, according to (Ehrhardt, 2008)

1.3   Classification in the Field of Tool Steels

Tool steels, according to DIN EN ISO 4957, are special steels used for processing (forming, die casting) and machining (cutting) workpieces, but also for handling equipment and measuring devices. They can be distinguished according to various criteria, such as chemical composition in unalloyed or alloyed tool steels, and according to the application temperature in cold or hot-work tool steels. Figure 1.2 shows an overview of groups of tool steels with a focus on their application possibilities. Plastic mold steels are also assigned to tool steels. However, they are not explicitly listed in the DIN EN ISO 4957 standard. These, as well as cold work and high-speed steels, are presented in separate *essentials*.

1.4   Designations

Material numbers
These are assigned by the European Steel Registry and consist of the main material group number (first number with a dot), the steel group numbers (second and third numbers), and the counting numbers (fourth and fifth numbers). For tool steels DIN EN 10027-2 divides the main material group 1 according to *steel group numbers* into:

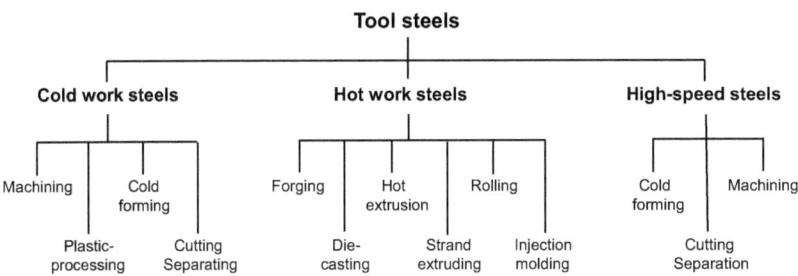

Fig. 1.2 Overview of the classification of tool steels

- *unalloyed tool steels: 1.15.. to 1.18..*
- *alloyed tool steels: 1.20.. to 1.28..*

The hot work tool steels can be assigned to the steel group numbers 1.20.. to 1.28.. as more or less alloyed tool steels. An exception are the unalloyed tool steels (carbon steels) such as 1.1625 (C80W2) and 1.1750 (C75W), which are used, for example for simple, smaller forging hammer dies, forging saddles, hot shears, riveting dies and similar applications. Two special materials are also assigned to the hot work materials as high-temperature-resistant alloys: 2.4668 (NiCr19Fe19Nb5Mo3) and 2.4973 (NiCr19CoMo), both used for tools for extruding heavy metals, such as dies, piercer plugs, and press plates, as well as for hot shear knives and sinter press tools.

Steel short names

For all the material numbers mentioned above, steel short names can be found, which are based on their steel's chemical compositions. They consist of main and additional symbols, which can be letters (e.g. chemical symbols) or numbers (for contents of alloying elements). These details differ for unalloyed, alloyed, and highly alloyed steels as well as for high-speed steels (Langehenke, 2007).

The *unalloyed tool steels* (carbon steels) are marked with the letter C for carbon, followed by the carbon content. The number specified for the carbon content is always multiplied by 100. To recognize the actual content, this number must be divided by 100. Additional symbols after the numerical value can provide information on special requirements, e.g., for the coating, treatment condition, or use.

Example: **C75W** (1.1750) – an unalloyed hot work tool steel with 75 / 100 = 0.75 mass-% carbon (**W** stands for welding wire).

For *low-alloyed tool steels*, the carbon content, also multiplied by the factor 100, is indicated in the short name at the first position. And, in contrast to the unalloyed steels, always without the letter C. This is followed by the chemical symbols for alloying elements and, in their respective order, their associated mass contents. It should be noted that these mass contents are always multiplied by different factors. These multipliers for the individual alloying elements are as follows:

- **Multiplier 4:** *Chromium (Cr), Cobalt (Co), Manganese (Mn), Nickel (Ni), Silicon (Si), Tungsten (W)*
- **Multiplier 10:** *Aluminum (Al), Beryllium (Be), Copper (Cu), Molybdenum (Mo), Niobium (Nb), Lead (Pb), Tantalum (Ta), Titanium (Ti), Vanadium (V), Zirconium (Zr)*
- **Multiplier 100:** *Carbon (C), Nitrogen (N), Phosphorus (P), Sulfur (S), Cerium (Ce)*
- **Multiplier 1000:** *Boron (B)*

To identify the actual alloy contents, the specified numbers in the steel short name must be divided by the associated multipliers.

Example: **55NiCrMoV7** (1.2714)—a nickel-alloyed hot work tool steel with 55 / 100 = 0.55 mass-% carbon and with 7 / Factor 4 = 1.75 mass-% nickel, this steel also contains chromium, molybdenum, and vanadium in comparatively lower contents.

The *high-alloyed tool steels* are marked with an **X** at the beginning of the short name. High-alloyed means that the average content of at least one alloying element is ≥5 mass-% (DIN EN 10027-1). The **X** is followed by the carbon content, again generally multiplied by a factor of 100, and the other alloying elements with their chemical symbols. The alloying elements are listed in decreasing order of their mass content. The mass fractions associated with each alloying element are then listed. However, these are not multiplied by a factor (typical for high-alloyed steels!).

Example: **X35CrMoV5-1-1** (1.2342)—a classic hot work tool steel with 35 / 100 = 0.35 mass-% carbon, approx. 5 mass-% chromium, 1 mass-% molybdenum, and 1 mass-% vanadium.

Brand names

In practice, manufacturers and dealers use their own designations, brand names, or protected trade names for their hot work tool steels. The steel designations for hot work tool steels produced by **DEW** (Deutsche Edelstahlwerke) provide information on special technologies and structures, e.g. **EFS** = extra fine structure, **Superclean** = remelted for highest purity, **Supercool** = special steel with very high thermal conductivity, or apply to special steels that are not standardized. **Thermodur®** is the general designation by DEW for hot work tool steel.

Examples:

Thermodur® 2329 corresponds to 1.2329 (46CrSiMoV7)

Thermodur® 2342 EFS corresponds to 1.2342 (X35CrMoV5-1-1)

Thermodur® 2999 Superclean (X45MoCrV5-3-1),

Thermodur® E 38 K Superclean and **Thermodur® E 40 K Superclean** correspond to non-standardized special steels.

Böhler (voestalpine High Performance Metals) as a steel producer also uses its own brand names with references to technology, such as: **ISODISC** = ingot casting, **ISOBLOC** = electro slag remelted, **VAR** = vacuum arc remelted

Examples:

Böhler W300 ISOBLOC® corresponds to 1.2343 (X37CrMoV5-1)

Böhler W303 ISODISC® corresponds to 1.2367 (X38CrMoV5-3)

Böhler W403 VMR® roughly corresponds to 1.2367 (X38CrMoV5-3)

UDDEHOLM (voestalpine High Performance Metals) classifies their produced hot work tool steels into the following groups with their own names:

- *conventional hot work tool steels:* **Formvar®, Orvar® 2 Microdized, Vidar™ 1, 1.2343** and **1.2344**
- *remelted hot work tool steels:* **Orvar® Supreme, Vidar™ 1 ESR, 1.2343 ESR**
- *Premium hot work tool steels:* **Dievar®, Qro® 90 Supreme, Unimax®, Vidar® Superior**

KIND&CO Edelstahlwerk Wiehl uses names like **RPU** for 1.2367 (X38CrMoV5-3), **USD** for 1.2344 (X40CrMoV5-1) and **USN** for 1.2343 (X37CrMoV5-1). The steel manufacturer **Friedr. Lohmann GmbH** labels the hot work tool steels in its delivery program with **LO-W**, e.g. **LO-W 2343** for 1.2343 (X37CrMoV5-1) or **LO-W 2367** for 1.2367 (X38CrMoV5-3).

Also **Dörrenberg Edelstahl** uses its own designations for hot work tool steels, e.g. **WP5** for 1.2343 (X37CrMoV5-1) and **A50** for 1.2714 (55NiCrMoV7).

And all other manufacturers worldwide, such as ArcelorMital, Nippon Koshuha Steel, Hitachi Metals (YXR 33), Aubert & Duval (SMR4), Villares Metals (VTM), Crucible (CPM1V), Sanyo Special Steel, Daido Steel, and Severstal trade hot work tool steel under their own names. This list is not exhaustive. It would otherwise exceed the scope of this brochure.

Designations according to international standards
Steels are classified with a **UNS** number (*abbreviation*: Unified Numbering System for Metals and Alloys), which is commonly used in the USA, such as **T20813** for the hot work tool steel **1.2344** (X40CrMoV5-1).

Based on country-specific standards, hot work tool steels can be found or compared on the steel market:

USA: **ASTM** (originally "American Society for Testing and Materials") as well as **AISI** (American Iron and Steel Institute). For example, the hot work tool steel mentioned above **1.2344** (X40CrMoV5-1) is designated as **H13** in ASTM A 681.

Germany: **DIN** (Deutsches Institut für Normung e. V.) **EN** (Europäische Normen). For example the above mentioned hot work tool steel H13 is designated as **1.2344** (X40CrMoV5-1) in DIN EN 4957 (Tool steels).

Japan: **JIS** (Japan Industrial Standard)
France: **AFNOR/NF** (Association Française de Normalisation)
United Kingdom: **BS** (British Standards)
Italy: **UNI** (Ente Nazionale Italiano di Unificazione)
China: **GB** (Guobiao), Chinese: National Standard)
Sweden: **SIS** (Swedish Institute of Standards)
Spain: **UNE** (Asociación Española de Normalización)
Poland: **PN** (from: Polish Committee for Normation).
Austria: **ÖNORM** (national österreichische **Norm**)   ·
Russia: **GOST** (Gosudarstvenny Standart)
Czech Republic: **CSN** (Czech national technical standard)

It should be noted in such a comparison that these are "equivalent", i.e. often only "comparable" hot work tool steels, which may differ slightly in the details of the chemical analysis. Figure 1.3 shows this using the example of grade 1.2344 (X40CrMoV5-1) with comparable assignable grades according to AISI (H13), UNS (T20813) and JIS (SKD61).

	Chemical composition (% by mass)							
	C	Si	Mn	P	S	Cr	Mo	V
Germany: X40CrMoV5-1	0,35-0,42	0,80-1,20	0,25-0,50	≤ 0,030	≤ 0,020	4,80-5,50	1,20-1,50	0,85-1,15
USA: AISI H13	0,32-0,45	0,80-1,20	0,20-0,50	≤ 0,030	≤ 0,030	4,75-5,50	1,10-1,75	0,80-1,20
UNS T20813	0,32-0,45	0,80-1,20	0,20-0,50	≤ 0,030	≤ 0,030	4,75-5,50	1,10-1,75	0,80-1,20
Japan: JIS SKD61	0,35-0,42	0,80-1,20	0,25-0,50	≤ 0,030	≤ 0,020	4,80-5,50	1,00-1,50	0,80-1,50

Fig. 1.3 Comparison of standards (chemical analyses) using the example of hot work tool steel 1.2344 (X40CrMoV5-1)

Chemical Compositions and Grades

2

2.1 Alloying Elements in Hot Work Tool Steels

The most important alloying elements in hot work tool steels are, in addition to carbon (C), chromium (Cr), tungsten (W), silicon (Si), nickel (Ni), molybdenum (Mo), manganese (Mn), vanadium (V), and cobalt (Co) and they have a significant influence on the transformation behavior during heat treatment and on the technological properties. These alloying elements are quantitatively coordinated so that the desired properties are achieved at the working temperature of the tools. The carbon content is responsible for the hardness increase (achievable hardness increase during hardening) of the steels, while the hardness penetration (penetration depth of the martensitic transformation) and the precipitation hardening (formation of secondary carbides during tempering) depend on the metallic alloying elements.

The following describes the influences of the alloying elements in more detail:

Carbon (C):
In addition to iron, tool steels have carbon as the most important alloying element. It is responsible for the formation of the martensitic microstructure and carbides with the elements chromium, tungsten, molybdenum, and vanadium. The mass fraction of carbon is adjusted to the mass fractions of these others elements. With a higher carbon content, the strength and hardenability of the steel increase, while ductility, forgeability, weldability and machinability decrease. The steel becomes more brittle.

© The Author(s), under exclusive license to Springer Fachmedien Wiesbaden GmbH, part of Springer Nature 2024
J. Schlegel and T. Schneiders, *Hot Work Tool Steel*,
https://doi.org/10.1007/978-3-658-43016-0_2

Chromium (Cr):

Chromium is a strong carbide former and improves the hardness penetration by lowering the critical cooling temperature. This is the material-dependent cooling rate that is necessary for the formation of the hard martensitic microstructure during hardening. The "upper critical cooling rate" represents the longest cooling duration or the lowest cooling rate to achieve 100% martensite, and the "lower critical cooling rate" represents the shortest cooling duration and thus the highest cooling rate at which martensite first appears. This allows tools with larger dimensions or cross-sections to be hardened.

In addition, chromium increases the high-temperature strength as well as the heat and corrosion resistance (steels with over 12 to 13 mass-% chromium are corrosion-resistant).

Tungsten (W):

Tungsten forms very hard carbides, improves toughness, and inhibits grain growth. At the same time, tungsten improves high-temperature strength, tempering resistance and wear resistance at high temperatures.

Silicon (Si):

Silicon is added to the steel melt in the form of the iron pre-alloy ferro-silicon for deoxidation (binding of the oxygen released during the cooling of the melt). Silicon has a solid-solution strengthening effect, increases scale resistance and quench hardness at higher contents, but also causes a decrease in toughness.

Nickel (Ni):

Nickel has a positive effect on the yield strength and toughness of steel. All transformation points of steel (temperatures at which phase transformations occur when exceeded or falling below) are lowered by nickel. Nickel alone makes steel only rust-resistant, but in austenitic steels combined with chromium, resistance is also achieved against oxidizing substances. Defined nickel contents lead to advantageous specific physical properties, e.g. very low thermal expansion (Invar steels).

Molybdenum (Mo):

Molybdenum is a strong carbide former and, like chromium, causes a decrease in the critical cooling rate. In addition, molybdenum contributes to the formation of special carbides, thus contributing to secondary hardness during tempering. Consequently, molybdenum, similarly to tungsten, favorably influences hardenability, temper brittleness, yield strength and tensile strength, as well as high-temperature strength. Scale resistance is reduced by molybdenum.

Manganese (Mn):
Manganese acts as a deoxidizer, i.e., it removes oxygen from steel and simultaneously binds sulfur. It dissolves in the steel matrix, does not form carbides and has a solid-solution strengthening effect (increases yield strength and tensile strength). The decrease in critical cooling rate caused by manganese in steel improves its hardenability. Manganese also positively affects forgeability and weldability, but negatively affects thermal expansion (increases it). Manganese also has a disadvantageous effect due to its tendency to form coarse grains and to increase the retained austenite content (Wendl, 1985). Retained austenite is the undesired austenitic phase in the desired martensitic structure that is usually present during conventional heat treatment by hardening and tempering. This means that the originally present austenite phase has not completely transformed into the martensite phase during quenching.

Vanadium (V):
Vanadium is also a strong carbide former. Like chromium and molybdenum, vanadium forms special carbides and is therefore very important for secondary hardening (Karagöz & Andrén, 1992). Wear resistance, high-temperature strength and tempering resistance are positively influenced.

Cobalt (Co):
Cobalt does not form carbides in steel, but it does improve tempering and wear resistance as well as high temperature strength. By adding cobalt, the highest hot hardness is achieved.

To ensure high temperature strength, but also thermal fatigue resistance and toughness of hot work tool steels, a microstructure of martensite with secondary carbide precipitations is required (Kulmburg, 1998), see Chap. 3: *Microstructure and properties.* Therefore, hot work tool steels without nickel, tungsten, and cobalt usually have alloy contents in the following ranges:

- *Carbon: 0.30 to 0.55 mass-%*
- *Chromium: 2.7 to approx. 5.5 mass-%*
- *Molybdenum: 1.1 to 3.2 mass-%*
- *Vanadium: 0.3 to 1.15 mass-%*

Nickel contents of 1.5 to 1.8 mass-% are found in e.g. the nickel-alloyed hot work tool steel 1.2714 (55NiCrMoV7). Tungsten-alloyed hot work tool steels can contain up to 9 mass-% tungsten, such as 1.2581 (X30WCrV9-3). And a cobalt-containing hot work tool steel, like 1.2661 (38CrCoW18-17-17), has 4.0 to 4.5 mass-% cobalt.

2.2   Types

A classification of hot work tool steels according to microstructure and grade is always related to the main application areas. These are die casting molds, extrusion dies, and forging tools, i.e. tools with cyclic loads in contact with workpieces that can have temperatures from 400 to 1200 °C. The contact duration between workpiece and tool can range from milliseconds (hammer dies) to minutes (extrusion). This results in different temperature loads and wear stresses, for which the hot work tool steels have been adapted in terms of their toughness and high temperature strength. Based on this, the following three groups can be distinguished:

• *Martensitic steels without or with low secondary hardness*, e.g. 1.2714 (55NiCr-MoV7).

This group offers only relatively low high temperature and creep strength.

• *Martensitic steels with pronounced secondary hardness*, e.g. 1.2344 (X40CrMoV5-1).

These steels with higher alloy contents of molybdenum and vanadium show higher hot strength due to hardening by special carbides.

• *Austenitic steels*, e.g. 1.2779 (X6NiCrTi26-15).

This third group contains the high temperature strength and scale-resistant austenitic hot work tool steels. They are used where very long contact durations at high workpiece temperatures occur, e.g., in extrusion of heavy metals. Austenitic hot work tool steels show higher high temperature strength at working temperatures above 650 °C than martensitic hot work tool steels.

By using state-of-the-art technologies during melting, remelting and heat treatment, hot work tool steels with particularly fine and homogeneous structure and highest purity can be produced today, contributing to the highest and most uniform tool life. This leads to another possibility of distinguishing hot work tool steels:

- *Conventional hot work tool steels for normal stress*
- *Remelted hot work tool steels and, if necessary, with special heat treatment for high stress*

Figure 2.1 shows an overview of the chemical analyses of the predominantly used hot work tool steels today, sorted by ascending material numbers.

Material number	Steel short name	Chemical composition (% by mass)											
		C	Si	Mn	P	S	Co	Cr	Mo	Ni	V	W	Others
Martensitic hot work tool steels													
1.1750	C75W	0,72-0,82	0,15	0,60-0,80	≤0,035	≤0,035	-	-	-	-	-	-	-
1.2082	X21Cr13	0,17-0,22	0,30-0,50	0,20-0,40	≤0,035	≤0,035	-	12,5-13,5	-	-	-	-	-
1.2083	X40Cr14	0,36-0,42	≤1,00	≤1,00	≤0,030	≤0,030	-	12,5-14,5	-	-	-	-	-
1.2309	65MnCrMo4	0,60-0,68	0,30-0,50	1,00-1,20	≤0,035	≤0,035	-	0,60-0,80	0,20-0,30	-	-	-	-
1.2311	40CrMnMo7	0,35-0,45	0,20-0,40	1,30-1,60	≤0,035	≤0,035	-	1,80-2,10	0,15-0,25	-	-	-	-
1.2312	40CrMnMoS8-6	0,35-0,45	0,20-0,40	1,40-1,60	≤0,030	0,05-0,10	-	1,80-2,00	0,15-0,25	-	-	-	-
1.2313	21CrMo10	0,16-0,23	0,20-0,40	0,20-0,40	≤0,025	≤0,025	-	2,30-2,60	0,30-0,40	-	-	-	-
1.2323	48CrMoV6-7	0,40-0,50	0,15-0,35	0,60-0,90	≤0,030	≤0,030	-	1,30-1,60	0,65-0,85	-	0,25-0,35	-	-
1.2329	46CrSiMoV7	0,43-0,48	0,60-0,75	0,65-0,85	≤0,030	≤0,030	-	1,65-1,85	0,25-0,35	0,45-0,60	0,17-0,22	-	-
1.2340	X36CrMoV5-1-1	0,32-0,40	≤0,50	≤0,50	≤0,010	≤0,010	-	4,60-5,40	1,10-1,60	≤0,30	0,35-0,60	-	-
1.2342	X35CrMoV5-1-1	0,30-0,40	0,70-1,20	0,40-0,60	≤0,030	≤0,030	-	4,50-5,50	1,00-1,20	-	0,80-1,00	-	-
1.2343	X37CrMoV5-1	0,33-0,41	0,80-1,20	0,25-0,50	≤0,030	≤0,020	-	4,80-5,50	1,10-1,50	-	0,30-0,50	-	-
1.2344	X40CrMoV5-1	0,35-0,42	0,80-1,20	0,25-0,50	≤0,030	≤0,020	-	4,80-5,50	1,20-1,50	-	0,85-1,15	-	-
1.2345	X50CrMoV5-1	0,40-0,53	0,80-1,10	0,20-0,40	≤0,030	≤0,020	-	4,80-5,20	1,25-1,45	-	0,80-1,00	-	-
1.2355	50CrMoV13-15	0,45-0,55	0,20-0,80	0,80-0,90	≤0,030	≤0,020	-	3,00-3,50	1,30-1,70	-	0,15-0,35	-	-
1.2357	50CrMoV13-14	0,45-0,55	0,20-0,50	0,50-0,80	≤0,030	≤0,020	-	3,00-3,60	1,20-1,60	-	0,05-0,25	-	-
1.2360	X48CrMoV8-1-1	0,40-0,50	0,70-0,90	0,35-0,45	≤0,020	≤0,005	-	7,30-7,80	1,30-1,50	-	1,30-1,50	-	-
1.2362	X63CrMoV5-1	0,60-0,65	1,00-1,20	0,30-0,50	≤0,035	≤0,035	-	5,00-5,50	1,00-1,30	-	0,25-0,35	-	-
1.2365	32CrMoV18-28	0,28-0,35	0,10-0,40	0,15-0,45	≤0,030	≤0,020	-	2,70-3,20	2,50-3,00	-	0,40-0,70	-	-
1.2367	X38CrMoV5-3	0,35-0,40	0,30-0,50	0,30-0,50	≤0,030	≤0,020	-	4,80-5,20	2,70-3,20	-	0,40-0,60	-	-
1.2564	30WCrV15-1	0,25-0,35	0,80-1,10	0,30-0,50	≤0,035	≤0,035	-	0,90-1,20	-	-	0,15-0,20	1,70-2,20	-
1.2567	30WCrV17-2	0,25-0,35	0,15-0,30	0,20-0,40	≤0,030	≤0,030	-	2,20-2,50	-	-	0,50-0,70	4,00-4,50	-
1.2581	X30WCrV9-3	0,25-0,35	0,10-0,40	0,15-0,45	≤0,030	≤0,020	-	2,50-3,20	-	-	0,30-0,50	8,50-9,50	-
1.2603	45CrVMoW5-8	0,40-0,50	0,50-0,70	0,30-0,50	≤0,035	≤0,035	-	1,30-1,60	0,40-0,60	-	0,75-0,90	0,40-0,60	-
1.2605	X35CrWMoV5	0,32-0,40	0,80-1,20	0,20-0,50	≤0,030	≤0,020	-	4,75-5,50	1,25-1,60	-	0,20-0,50	1,10-1,60	-
1.2606	X37CrMoW5-1	0,32-0,40	0,90-1,20	0,30-0,60	≤0,035	≤0,035	-	4,50-5,50	1,30-1,60	-	0,15-0,40	1,20-1,40	-
1.2622	X60WCrMoV9-4	0,55-0,65	0,20-0,40	0,20-0,40	≤0,035	≤0,035	-	3,70-4,20	0,80-1,00	-	0,60-0,80	8,50-9,50	-
1.2661	38CrCoWV18-17-17	0,35-0,45	0,15-0,50	0,20-0,50	≤0,030	≤0,020	4,00-4,50	4,00-4,70	0,30-0,50	-	1,70-2,10	3,80-4,50	-
1.2662	30WCrCoV9-3	0,27-0,32	0,15-0,30	0,20-0,40	≤0,035	≤0,035	1,80-2,30	2,20-2,50	-	-	0,20-0,30	8,00-9,00	-
1.2678	X45CoCrWV5-5-5	0,40-0,50	0,35-0,50	0,30-0,50	≤0,025	≤0,025	4,00-5,00	4,00-5,00	0,40-0,50	-	1,80-2,10	4,00-5,00	-
1.2709	X3NiCoMoTi18-9-5	≤0,03	≤0,10	≤0,10	≤0,010	≤0,010	8,50-10,0	≤0,25	4,50-5,20	17,0-19,0	-	-	Ti 0,8-1,2
1.2711	54NiCrMoV6	0,50-0,60	0,15-0,35	0,50-0,80	≤0,025	≤0,025	-	0,60-0,80	0,25-0,35	1,50-1,80	0,07-0,12	-	-
1.2713	55NiCrMoV6	0,50-0,60	0,10-0,40	0,65-0,95	≤0,030	≤0,030	-	0,60-0,80	0,25-0,35	1,50-1,80	0,07-0,12	-	-
1.2714	55NiCrMoV7	0,50-0,60	0,10-0,40	0,60-0,90	≤0,030	≤0,030	-	0,80-1,20	0,35-0,55	1,50-1,80	0,05-0,15	-	-
1.2726	26NiCrMoV5	0,22-0,30	0,20-0,50	0,20-0,40	≤0,030	≤0,030	-	0,60-0,90	0,20-0,40	1,30-1,60	0,15-0,20	-	-
1.2738	40CrMnNiMo8-6-4	0,35-0,45	0,20-0,40	1,30-1,60	≤0,035	≤0,035	-	1,80-2,10	0,15-0,25	0,90-1,20	-	-	-
1.2740	28NiCrMoV10	0,24-0,32	0,30-0,50	0,20-0,40	≤0,030	≤0,030	-	0,60-0,70	0,50-0,70	2,30-2,60	0,25-0,32	-	-
1.2743	60NiCrMoV12-4	0,55-0,60	0,30-0,50	0,50-0,80	≤0,035	≤0,035	-	1,00-1,30	0,30-0,40	2,70-3,00	0,07-0,12	-	-
1.2744	57NiCrMoV7-7	0,50-0,60	0,15-0,35	0,60-0,80	≤0,035	≤0,035	-	0,90-1,20	0,70-0,90	1,50-1,80	0,07-0,12	-	-
1.2747	28NiMo17	0,24-0,31	0,15-0,35	0,20-0,40	≤0,030	≤0,030	-	0,30-0,50	1,15-1,25	4,20-4,70	0,15-0,20	-	-
1.2766	35NiCrMo16	0,32-0,38	0,15-0,30	0,40-0,60	≤0,035	≤0,035	-	1,20-1,50	0,20-0,40	3,80-4,30	-	-	-
1.2767	45NiCrMo16	0,40-0,50	0,10-0,40	0,20-0,50	≤0,030	≤0,030	-	1,20-1,50	0,15-0,35	3,80-4,30	-	-	-
1.2787	X23CrNi17	0,10-0,25	≤1,00	≤1,00	≤0,030	≤0,030	-	15,5-18,0	-	1,00-2,50	-	-	-
1.2885	X32CrMoCoV3-3-3	0,28-0,35	0,10-0,40	0,15-0,45	≤0,030	≤0,030	2,50-3,00	2,70-3,20	2,60-3,00	-	0,40-0,70	-	-
1.2886	X15CrCoMoV10-10-5	0,13-0,18	0,15-0,25	0,15-0,25	-	-	9,50-10,5	9,50-10,5	4,90-5,20	-	0,45-0,55	-	-
1.2888	X20CoCrWMo10-9	0,17-0,23	0,15-0,35	0,40-0,60	≤0,035	≤0,035	9,50-10,5	9,00-10,0	1,80-2,20	-	-	5,00-6,00	-
1.2889	X45CoCrMoV5-5-3	0,40-0,50	0,30-0,50	0,20-0,50	≤0,025	≤0,025	4,00-5,00	4,00-5,00	2,80-3,30	-	1,80-2,10	-	-
1.2999	X45MoCrV5-3-1	~0,45	~0,30	~0,30	≤0,030	≤0,030	-	~3,0	~5,0	-	~1,0	-	-
Austenitic hot work tool steels													
1.2731	X50NiCrWV13-13	0,45-0,55	1,20-1,50	0,60-0,80	≤0,035	≤0,035	-	12,0-14,0	-	12,5-13,5	0,30-1,00	1,50-2,80	-
1.2779	X8NiCrTi26-15	≤0,08	≤0,50	≤0,50	≤0,035	≤0,035	-	13,5-16,0	1,00-1,50	24,0-27,0	0,10-0,50	-	Ti 1,9-2,3
1.2782	X16CrNiSi25-20	≤0,20	1,80-2,30	≤2,00	≤0,035	≤0,035	-	24,0-26,0	-	19,0-21,0	-	-	-
1.2786	X13NiCrSi36-16	≤0,15	1,50-2,00	≤2,00	≤0,035	≤0,035	-	15,0-17,0	-	34,0-37,0	-	-	-
Nickel-based alloys													
2.4668	NiCr19Fe19Nb5Mo3	0,02-0,08	≤0,35	≤0,35	≤0,015	≤0,015	≤1,00	17,0-21,0	2,80-3,30	50,0-55,0	-	-	Fe rest/bal. Nb/Ta 4,7-5,5 Ti 2,8-3,3
2.4973	NiCr19CoMo	≤0,12	≤0,50	≤0,10	-	-	10,0-12,0	18,0-20,0	9,00-10,5	rest/bal.	-	-	Fe ≤5,00 Ti 2,8-3,3 Al 1,4-1,8

Fig. 2.1 Comparison of chemical analyses of hot work tool steels

Microstructure and Properties   3

The properties of a steel are determined by its microstructure formation, which in turn is influenced by the chemical composition and processing, including heat treatment. This applies generally to all steels and thus also to hot work tool steels. Figure 3.1 shows this relationship for tool steels with their specific requirements.

The heat-treatable hot work tool steels are usually delivered as semi-finished products or cuttings in the soft annealed state, which is easy to machine (max. 235 HB), with a ferritic matrix structure containing spherically shaped carbides (Schruff, 1989). Figure 3.2 shows this for the example of the hot work tool steel **1.2343** (X37CrMoV5-1) with carbides precipitated and spherically shaped in the soft ferritic matrix by soft annealing with slow cooling.

This soft annealed microstructure is well suited as a starting point for mechanical processing to produce the desired tools. For their use, a microstructure of martensite with secondary carbide precipitations is necessary to ensure high temperature strength, as well as thermal fatigue resistance and toughness. This is achieved by targeted heat treatment (hardening and tempering in the highest possible temperature range, this combination is called QT—*quenching and tempering*) of the tools after machining. Figure 3.3 shows a micrograph of the microstructure of a quenched and tempered, martensitic hot work tool steel using the example of **1.2343** (X37CrMoV5-1). The martensitic, needle-like structure of the hardened microstructure is clearly visible. Due to their very small size, the carbide precipitations are difficult to discern.

The tools for hot forming and pressure die casting made of hot work tool steels are subject to complex mechanical, thermal, chemical, and tribological stresses that occur cyclically. Figure 3.4 shows these complex stresses using the example of a forging die.

© The Author(s), under exclusive license to Springer Fachmedien Wiesbaden GmbH, part of Springer Nature 2024
J. Schlegel and T. Schneiders, *Hot Work Tool Steel*,
https://doi.org/10.1007/978-3-658-43016-0_3

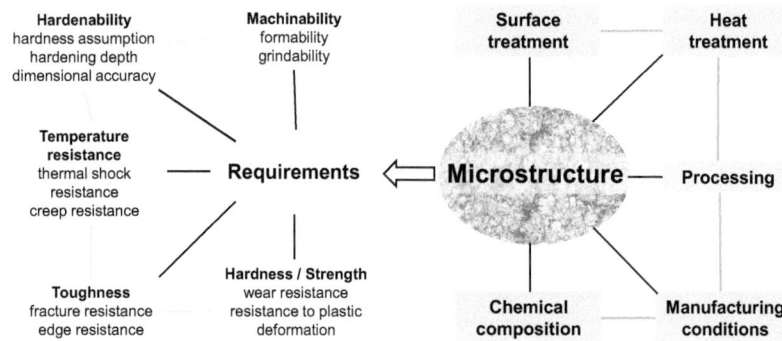

Fig. 3.1 Influence of microstructure formation and properties of tool steels according to (Schruff, 2002)

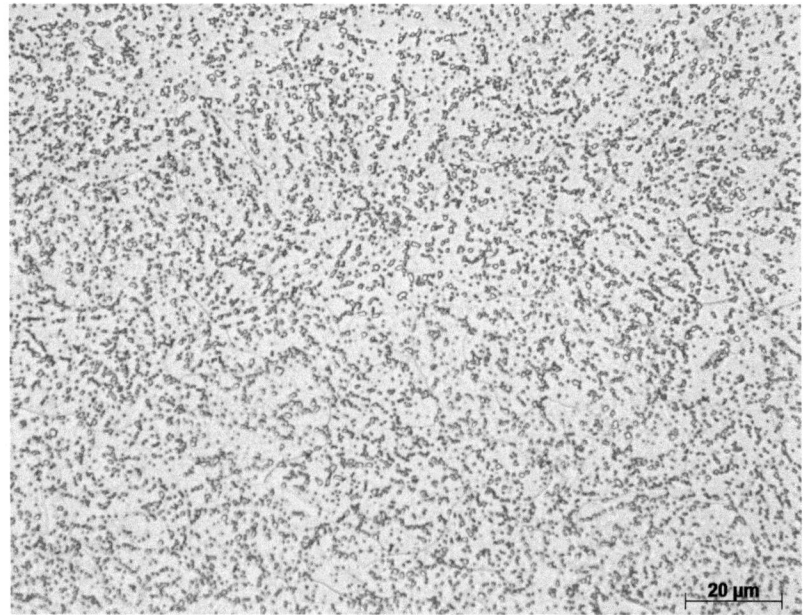

Fig. 3.2 Soft annealed microstructure of the hot work tool steel **1.2343** (X37CrMoV5-1) with clearly visible spherically shaped carbides in the ferritic matrix (micrograph: Deutsche Edelstahlwerke Specialty Steel GmbH & Co.KG)

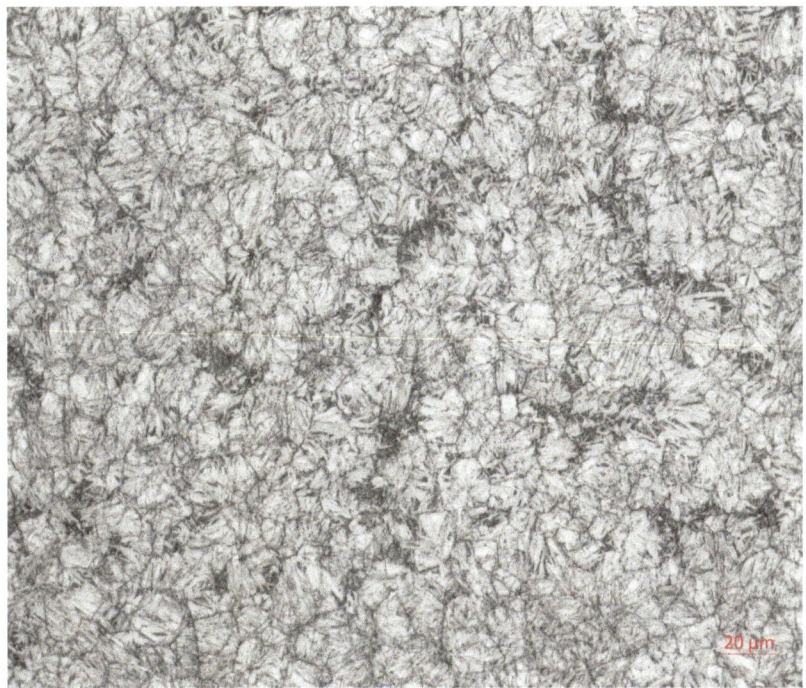

Fig. 3.3 Cross-sections of the hot work tool steel 1.2343 (X37CrMoV5-1): bar, hot rolled and quenched and tempered, 500× magnification (micrograph: BGH Edelstahl Lugau GmbH)

The following **properties of hot work tool steels** are required:

- *high microstructural uniformity*
- *high hot strength and toughness*
- *high tempering resistance*
- *high hot wear resistance*
- *high scale resistance*
- *high resistance to temperature changes (thermal shock resistance)*
- *high thermal conductivity*
- *good hardenability*
- *good machinability and coatability*
- *good dimensional stability*

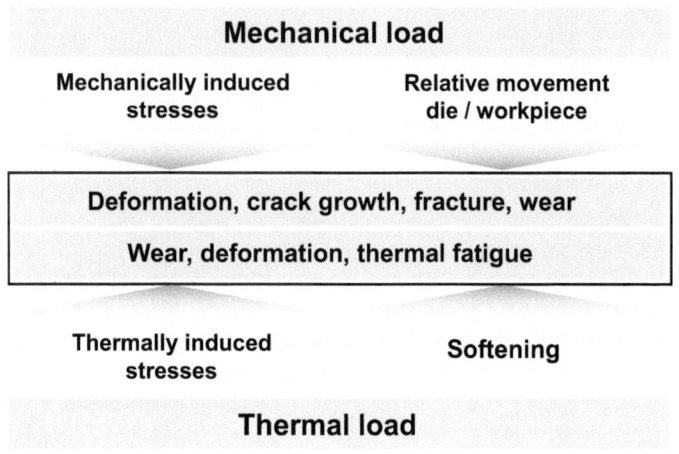

Fig. 3.4 Illustration of the complex stress of a forging die

- *low tendency to warp*
- *low tendency to stick*
- *high resistance to erosion, high-temperature corrosion and oxidation*

Hot strength

Hot strength describes the ability of a material to withstand loads (mechanical stresses) even at elevated temperatures and to "endure" these without permanent deformation.

The predominantly used martensitic hot work tool steels have tensile strengths R_m in the range of 1200 to over 2300 N/mm² at room temperature (König & Klocke, 2006). Depending on the test temperatures, the hot work tool steels show characteristic curves of hot strength depending on the alloy composition. In the temperature range around 400 °C, the hot strengths R_m are still about 1200 to 1400 N/mm². Figure 3.5 shows curves of the hot strength (R_m and $R_{p0,2}$) as well as the reduction in area (Z) for the hot work tool steel **1.2343** (X37CrMoV5-1) as an example.

Tempering resistance

Tempering resistance characterizes the resistance of a material to softening at increasing temperatures. The characteristic tempering resistance of a hot work

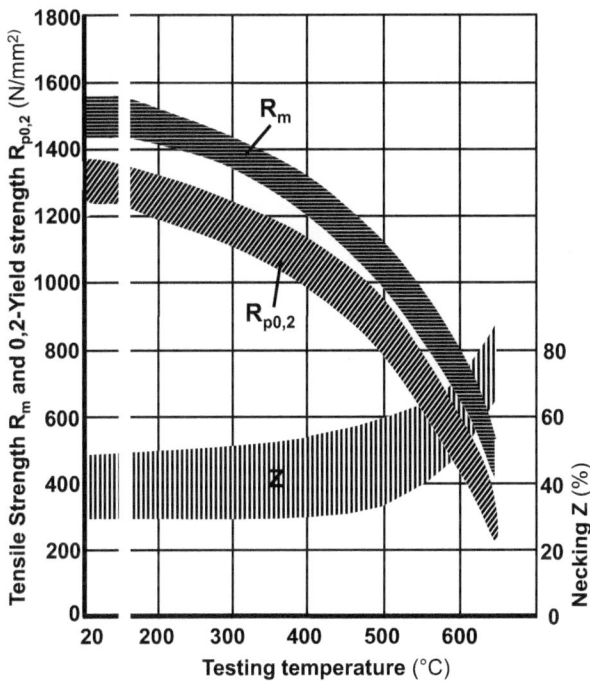

Fig. 3.5 Curves of the hot strength of the hot work tool steel 1.2343 (X37CrMoV5-1) according to (Schruff, 2002)

tool steel, which also favors its application range up to more than 600 °C, can be seen from the comparison of its tempering curve with the tempering curves of a cold work tool steel and a high-speed steel. This comparison is shown in Fig. 3.6.

If, for example, during die forging, the tempering effect of quenching and tempering is exceeded at too high temperatures, the surface layer of the tool softens. Die edges in forging dies absorb heat from both sides, so they reach a higher temperature than flat die surfaces. In addition, such edges are usually mechanically more highly stressed, so that deformation and softening mutually reinforce each other. In combination with the tribological load, wear is increased in these areas (Berns, 2004).

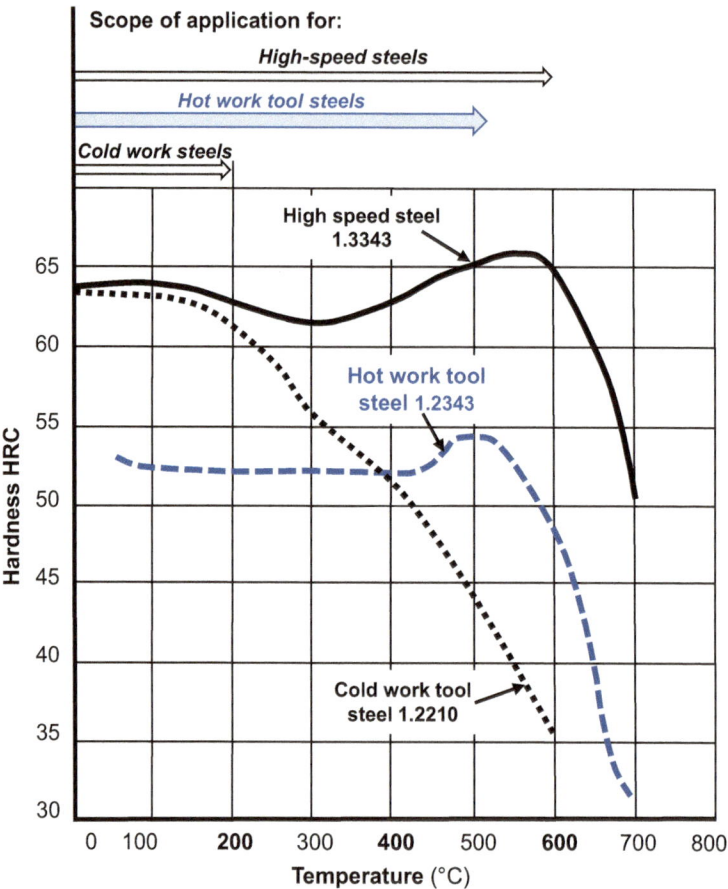

Fig. 3.6 Comparison of the tempering curves of a hot work tool steel, a cold work tool steel and a high-speed steel

Thermal Fatigue Resistance

The service life of a hot work tool is significantly determined by its resistance to constant temperature changes. For example, the main failure cause of die casting molds, with a share of 80%, is the formation of thermal shock cracks or thermal fatigue (Schruff, 2003). Under production conditions, the tool and the hot work-pieces are in contact, causing the tool surface to be subjected to cyclic thermal and mechanical stress. The surface layer of the tool will heat up faster during this cyclic thermal load than the bulk material. This creates a steep temperature gradient, so that stresses can build up in the surface layer (Schneiders, 2005). These thermally induced stresses cause the formation of microcracks, similar to fatigue. A delay in the formation of microcracks (incipient cracks) can be achieved by using a hot work tool steel with higher heat resistance and ductility. The further service life of the tool depends crucially on the progression of the formed cracks (crack propagation). For this reason, a high fracture toughness (resistance to crack propagation) of the material is also advantageous.

Hot Wear Resistance
According to DIN 50320, wear is defined as *"progressive material loss from the surface of a solid body caused by mechanical factors, i.e. by contact movements and relative movements of a solid, liquid, or gaseous body"*. Responsible for such mass loss are the four wear mechanisms adhesion (adherence at the contact surfaces), abrasion (superficial removal by friction), tribochemical reaction and surface fatigue (Macherauch & Zoch, 2011). The main wear mechanisms during forming are adhesion and abrasion. During contact between the tool and workpiece, adhesion occurs at roughness peaks. This is caused by local sticking and welding due to molecular and atomic interactions. The relative movement between the tool and workpiece separates these connections. This can occur at the original contact surfaces or in the near-surface areas of the involved partners, resulting in material removal. Abrasion is caused by the penetration of hard particles into a softer surface. In hot work tools exposed to higher temperatures during operation, scale primarily appears as an abrasive. As a result of the relative movements between the tool and workpiece, the micro-mechanisms of micro-ploughing, micro-cutting, micro-fatigue and micro-cracking occur. The tool surface shows grooves, chipped areas, breakouts and cracks after abrasive stress, leading to material loss. Tribochemical reactions are chemical interactions occurring between the tool, workpiece, lubricant, and surrounding medium. Surface fatigue is caused by material fatigue, crack formation and material removal due to mechanical and thermal stress (Schneiders, 2005).

High-Temperature Corrosion
The service life of the tool depends on the resistance or durability of a tool material against the mentioned wear mechanisms during its application. In addition, the surfaces of hot work tool steels in operation at elevated temperatures are also exposed to the action of various media such as air, lubricants, coolants and workpiece material. Above 570 °C, iron oxidizes to wustite (FeO). This would

normally lead to a thick, rapidly growing scale layer, which would cause material loss through flaking. In hot work tool steels, this is prevented by the increased chromium content. This causes the formation of a thin, adherent oxide layer (Berns, 1993). Therefore, oxidative removal is usually not the main focus in hot work tool steels (Berns, 2004).

Graphite-containing coolants have a carburizing effect, but due to the low solubility of carbon in the body-centered cubic lattice, this effect can be neglected when hot forming tools made of hot work tool steel. However, if the structure of the tool's surface layer is transformed (austenitized) during operation at excessive temperatures, decarburization of the surface can occur due to oxidation of the carbon (Berns, 2004). This leads to a reduction in surface hardness and thus increased wear.

In die casting of light and heavy metals, significant material loss can occur on the tool. The reasons for this include the dissolution of the tool material in the liquid metal, e.g. aluminum (Persson et al., 2002).

Thermal conductivity
This property of a material describes its ability to conduct heat well, or in other words, how fast heat spreads from one point through the material (forming tool). High thermal conductivity is important for hot forming and die casting tools to quickly dissipate temperature differences and avoid damaging temperature peaks on the tool surface, deformations, and internal stresses. Furthermore, high thermal conductivity can contribute to a reduction in cycle time, e.g. in die casting or plastic processing.

Toughness
In the context of the mentioned properties such as high temperature strength, tempering, and scale resistance in connection with the stresses during the use of tools, the hot toughness of the used hot work tool steels plays a special role concerning the achievable tool life (see Fig. 1.1). In general, toughness is understood as the resistance of a material to fracture or crack propagation (Issler et al., 2003). This is usually determined in the impact bending and notched impact bending test as energy of rupture related to the nominal cross-section of the sample. Toughness is thus the ability of a material to absorb mechanical energy during plastic deformation without breaking. The opposite of toughness is brittleness. And toughness should not be confused with ductility. Ductility describes the property of a material to permanently plastically deform under load before breaking, determinable, for example, with the tensile test (Gottstein, 2014).

Property	Definition	Effect / Benefits
Hot toughness	Resistance of a material to cracking or crack propagation.	Hot toughness reduces the risk of cracking and crack propagation, important for tools with deep engravings, cross-sectional transitions and edges. Stress peaks are reduced and good dimensional stability of the tools is achieved.
High temperature strength	The ability of a material to absorb loads (mechanical stresses) without permanent deformations, even at elevated service temperatures.	Sufficient high-temperature strength, i.e. "bearable" loads even at high temperatures, provides safety against deformation and wear of the tools.
Tempering resistance	Resistance of a material to softening at increasing temperatures.	A high tempering resistance of the hot work steels leads to sufficient working hardness even at elevated temperatures.
Hot wear resistance	Resistance to progressive material loss on the tool surface caused by mechanical effects (adhesion, abrasion, tribochemical reactions and contact fatigue).	A high resistance to heat wear reduces the risk of erosion, i.e. signs of wear on the mold contours of the tools.
Thermal shock resistance	The ability of a material to withstand constantly repeating rapid temperature changes occurring during continuous operation.	Temperature changes are particularly harsh stresses. The higher the thermal shock resistance of the hot work tool steel, the lower the risk of stress cracks and thus damage to the surface of the tool.
Thermal conductivity	The thermal conductivity of a material indicates how well it conducts heat or how well it is suitable for thermal insulation.	A high thermal conductivity reduces the temperature differences and thus stresses in the tool. Damaging temperature peaks on the tool surface and deformations are avoided.

Fig. 3.7 Overview of important properties of hot work tool steels and their effects or benefits (Based on a presentation from the company information on hot work tool steel by voestalpine Böhler Edelstahl GmbH & Co KG, 2018)

Toughness allows the dissipation of stress peaks occurring during operation due to mechanical or thermal overloading, and delays crack formation and propagation due to thermal fatigue, or unsuitable tool cross-sections. This is particularly important when using hot work tool steels for forming tools and die casting molds, as heat cracks and stress cracks occur especially in tools with deep engravings, at cross-sectional transitions and along edges. Since the plastic deformation capacity of tool steels is only desired in combination with a simultaneously high elastic limit, a high elastic limit is also the essential criterion for the fracture safety of these tool steels (Kulmburg et al., 1994). In addition, high hot strength and tempering resistance are prerequisites for good dimensional stability of the tools.

Figure 3.7 provides an overview of the important properties of hot work tool steels and their effects or benefits.

Manufacturing 4

The production of hot work tool steels and the tools made from them includes the melting or powder metallurgical production, including secondary metallurgical treatments, further processing into semi-finished products and finished products (tools), heat treatments, and possibly additional final surface treatments. Whether conventional or remelted hot work tool steels, manufacturers use adapted and very different production technologies for each quality.

4.1 Melting Metallurgical Production

Alloyed tool steels, including hot work tool steels, are nowadays produced in electric steel plants from pure-grade scrap (Ernst, 2009). Modern electric steel plants operate with electric arc furnaces with batch sizes up to 200 tons. In the electric arc furnace (EAF), the current (usually three-phase) forms an electric arc between the current-carrying graphite electrodes and the scrap input. This arc melts the scrap through thermal radiation. The raw steel is then cast into a preheated ladle. In downstream secondary metallurgical plants, further "refining" of the still liquid raw steel is carried out: alloying of certain alloying elements, homogenization of the melt, reduction of carbon and sulfur content, setting of the casting temperature. For high-quality, alloyed hot work tool steels, vacuum treatment in the VOD converter (**v**acuum-**o**xygen-**d**ecarburization—decarburization under vacuum with oxygen) is predominantly used. After completion of this fine treatment, usually also called "ladle metallurgy" or "secondary metallurgical treatment" (Burghardt & Neuhof, 1982), the finished steel melt is cast as ingot casting or pre-block continuous casting.

For particularly high requirements regarding purity and homogeneity (reduction of segregations in the cast structure), remelting may be necessary. Electroslag remelting plants (ESR) or vacuum arc remelting furnaces (VAR) are used to subject the already melted, secondary metallurgically treated and cast steel to a further cleaning and refining process.

In the ESR process, remelting takes place in a reactive slag. This reduces undesired accompanying elements and non-metallic inclusions, resulting in an improved purity level without changing the basic composition of the steel. The controlled axially solidification leads to dense, very homogeneous ingots (N.N., 1994).

Remelting in the electric arc furnace under vacuum, also called VAR (vacuum-arc-remelting), also leads to an improvement in purity level. Oxidation of the melted material is prevented, and additionally, the content of dissolved gases such as oxygen, hydrogen, and nitrogen (Trenkler & Kreiger, 1988) and the content of unwanted trace elements can be reduced (N.N., 1994). Although the sulfur content is not significantly reduced during VAR, the sulfides are more finely distributed (Trenkler & Kreiger, 1988).

Comparing the remelting processes ESR and VAR, ESR offers lower remelting costs, more intensive desulfurization, higher flexibility of ingots weights due to the possibility of rapid electrode changes, and higher quality of the ingot surface. Ingots remelted under vacuum (VAR) have minimal gas contents, reduced contents of trace elements such as lead, bismuth and tellurium, lower micro-segregation levels in the block center, and more precise adjustment of the chemical composition (N.N., 1994).

Using the example of hot work tool steel 1.2367 (X38CrMoV5-3), the effects of remelting on purity and toughness can be illustrated. The steel produced via ingot casting achieves K0 values according to DIN 50602 between 10 and 50 in terms of purity. The electroslag remelted steel has K0 values between 5 and 20, while a vacuum remelted variant of this steel shows values below 6. With this increase in purity and the simultaneous reduction of segregations, the toughness is increased. Accordingly, the VAR variant achieves the highest toughness values of up to 500 J/cm^2 in the impact bending test. The ESR steel still achieves more than 400 J/cm^2, while the ingot cast variant has impact bending toughness values of only about 100 J/cm^2 (Jung, 2003).

After casting and any remelting, the hot forming (forging, rolling) of the cast ingots into semi-finished products round, square or flat takes place. During this hot forming and the associated heat input, remaining segregations are reduced, and the precipitated carbides are largely dissolved again (Gümpel, 1983). Figure 4.1 shows a simplified complete process route of the melt metallurgical

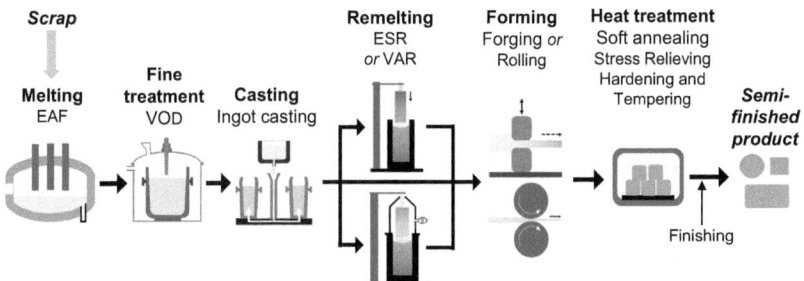

Fig. 4.1   Process route of the melt metallurgical production of hot work tool steels and their further processing into tools

production of hot work tool steels, including further processing into semi-finished products and heat treatment.

4.2   Powder Metallurgical Production

Since the 1960s, tool steels have been produced powder metallurgically on an industrial scale (Grinder, 1999), with hot work tool steels being produced in this way only since the 1980s (Bayer & Seilstorfer, 1984). Compared to melt metallurgical production, the more cost-intensive powder metallurgical production offers some advantages. For example, the hot formability of melt metallurgically produced tool steels decreases with increasing alloy content. In contrast, there are no formability limits for powder metallurgically produced blocks for much larger alloy ranges (Wilmes, 1990). The reason for this is the homogeneous, segregation-minimized, fine microstructure with evenly distributed, small carbides in the micrometer range.

The powder metallurgical production route initially began with melting, powder atomization under inert gas, encapsulation of the powder, and subsequent hot isostatic pressing (compaction). Later, other powder metallurgical process routes were added, such as vacuum or liquid phase sintering (Grinder, 1999). Ultimately, the **h**ot **i**sostatic **p**ressing (**HIP**) technique has also become established for the production of hot work tool steels, albeit to a limited extent. Only this process route will be mentioned below, i.e. the three main steps of metal powder production, shaping/compacting of the powder (HIP process), and heat treatment/sintering.

The powder production begins with the generation of a melt in the induction furnace. The desired chemical composition is achieved by using scrap in combination with unalloyed and low-alloyed steels, alloying elements, and powder metallurgically produced process scrap, controlled and, if necessary, corrected (Bockholt, 2002). The finished steel melt is directed into a casting distributor, where the deposition of non-metallic slags takes place, thus improving the purity level. At the bottom of the distributor, a nozzle is attached through which the melt flows out and is atomized using nitrogen. The produced powder has a spherical shape and can be immediately filled into capsules, compacted to a preform close to the theoretical density and simultaneously sintered. This is done by hot isostatic pressing (HIP) in a heatable pressure chamber under protective gas (argon) at temperatures around 1150 °C and pressures around 100 MPa. Figure 4.2 shows a simplified version of this special HIP process.

The powder compaction is based on diffusion processes between the powder grains (surface, grain boundary and lattice diffusion) as well as on plastic deformation. The hot work tool steel ingots produced by the HIP process are forged or hot-rolled into semi-finished products. For further processing to the desired final products, such as die-casting molds or forging dies, machining processes, thermal

Hot isostatic pressing (HIP)

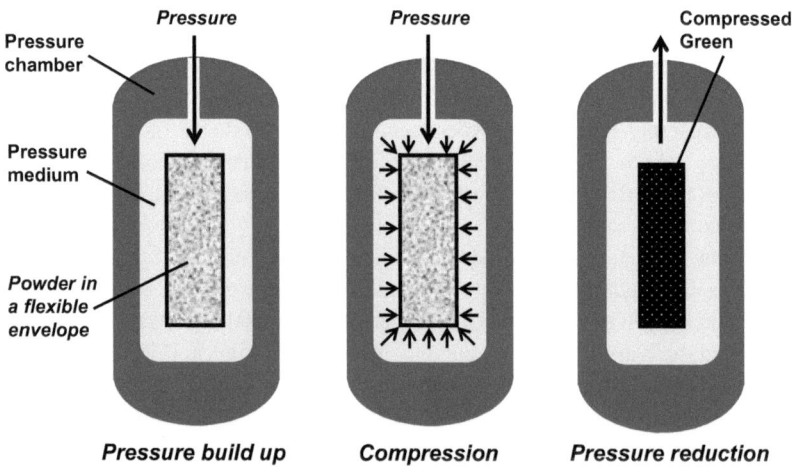

Fig. 4.2   Principle of the hot isostatic pressing process (HIP process)

treatments (hardening) and surface treatments (coatings) are used. While powder metallurgically produced cold work and high-speed steels have now been able to establish themselves on the market, the application of powder metallurgical hot work tool steels is limited to a few special applications (Schneiders, 2005).

4.3 Further Processing

The metallurgically produced and formed semi-finished product made of hot work tool steel is further processed using manufacturing techniques to create a hot work tool, i.e. a product precisely defined in terms of shape, dimensions, dimensional tolerances, surface quality, and mechanical properties. Various manufacturing techniques are usually used for this purpose. These include mechanical processes such as turning, milling, planing/shaping, drilling and grinding, as well as thermal processes such as die-sinking and wire electrical discharge machining to create the desired contours in a die-casting mold, extrusion die or forging die. With such processing, a diverse range of stresses on the tools begins, which can also lead to damage. For example, during mechanical processing, the surface layer of the tool is heated, plastically deformed and subjected to residual stresses. However, due to the high tempering temperature of hot work tools, softening due to increased temperatures during machining is not expected. A stress-relief treatment to reduce hardening and residual stresses is always recommended. Also, a low surface roughness should always be maintained, as machining grooves are preferred starting points for fatigue cracks (Berns, 2004).

In the case of electrical discharge machining, the tool surface is melted drop by drop through electrical discharges. The remaining surface layer consists of an outer, melted and re-solidified layer and an inner heat-affected zone. The melted surface layer can sometimes be crack-prone. To exclude damage during operation, it is recommended to mechanically remove this melt zone and also to stress-relieve the tool (Becker & Kiel, 1983).

4.4 Heat Treatment

During the production and after completion of hot work tools, heat treatments are carried out. These include intermediate heat treatments such as the annealing processes *diffusion annealing, soft annealing* and *stress-relief annealing* as well as final heat treatments such as *quenching and tempering* (hardening and tempering). The influencing factors are heating (heating and holding time), temperature, atmosphere (air, vacuum, protective gas) and cooling (cooling rate). In different combinations and sequences, these cause changes in the steel microstructure (precipitations and phase transformations, changes in their proportions, their arrangement, shape, and composition), thereby adjusting the desired properties (Weißbach, 2007). Figure 4.3 shows the temperature ranges of the mentioned heat treatment processes for hot work tool steels, plotted in the simplified iron-carbon diagram.

Based on these temperature ranges of the different heat treatment processes, Fig. 4.4 comparatively shows the associated time-temperature curves.

Not only the certain hardnesses on the finished hot work tool are adjusted by heat treatment, but also other mechanical properties, such as hot toughness and thermal shock resistance. And considering the often very large-sized tools, it quickly becomes clear that very special requirements are placed on the heat treatment processes.

Soft annealing

The delivery condition for hot work tool steels for the production of tools is usually the soft-annealed, easily machinable state. The so-called annealed hardnesses are in the range of 210 to approx. 320 HB, depending on the alloy. Soft annealing is usually carried out at temperatures in the range of 650 to 750 °C, always with temperature ranges of approx. 30 to 50 K. The holding times are usually more than 4 h, depending also on the workpiece size. Slow furnace cooling down to 500 °C is important, after which further air cooling can take place. This creates a structure in the hot work tool steel consisting of a ferritic matrix with spherically shaped carbides (Schruff, 1989), see example in Fig. 3.2.

Specific soft annealing parameters for individual hot work tool steels can be found in the data sheets in Chap. 6: *Material data sheets*.

Diffusion annealing

Diffusion annealing is used to reduce microstructural inhomogeneities (local differences in the chemical composition of steels caused by segregation). This is also where the common term homogenizing annealing originates from (https://en.wikipedia.org/wiki/Annealing_(materials_science)). With this heat treatment, the degree of segregation can be reduced and thus the toughness increased.

Since diffusion processes are strongly dependent on temperature and time, heating to very high temperatures (approx. 1100 and 1300 °C) is required, see Fig. 4.3. In addition, very long holding times, possibly up to 50 h, are necessary. The cooling must be carried out slowly. Such diffusion annealing is very complex and cost-intensive due to the high temperatures and long annealing times. It is

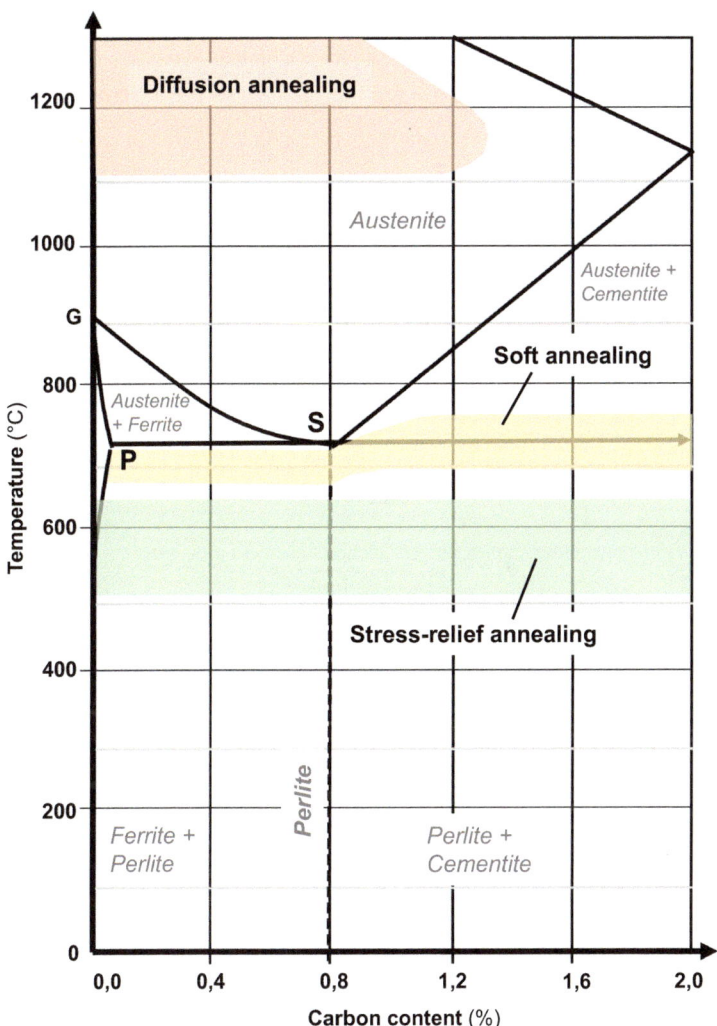

Fig. 4.3 Temperature ranges of the heat treatment types, shown in the simplified iron-carbon diagram

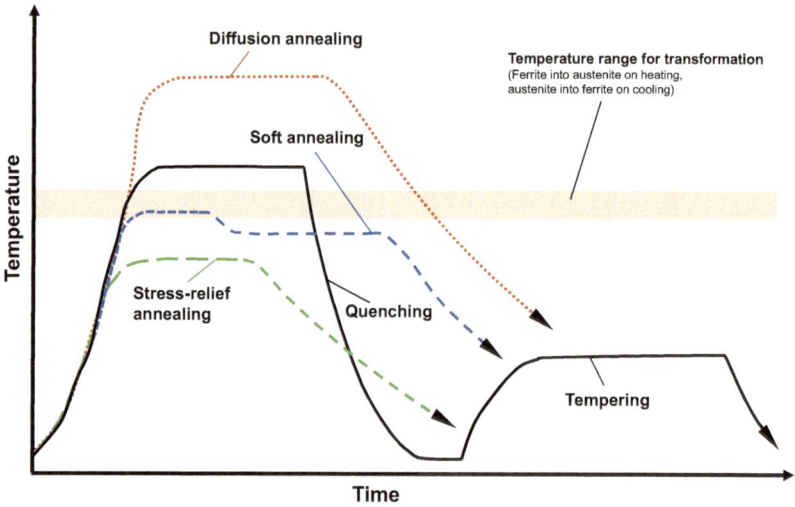

Fig. 4.4 Time-temperature curves for various heat treatment processes

mainly used when segregations in the structure are to be reduced in ingots. Due to the high scale losses, diffusion annealing is usually carried out before hot forming.

Quenching and tempering (hardening and tempering)
The desired performance characteristics of hot work tools are adjusted after processing by quenching and tempering. "Quenching and tempering" refers to the combination of hardening and tempering processes in the upper possible temperature range. These final heat treatments ensure high hardness and the necessary toughness at the same time, thus adjusting the desired performance characteristics.

Hardening
Hardening serves to increase the mechanical resistance (hardness, strength) of the steel by a targeted change in the microstructure through transformation (https://en.wikipedia.org/wiki/Hardening_(metallurgy)). This hardening process works because the steel to be hardened shows a phase transformation from ferrite to austenite or austenite to ferrite during heating and cooling. However, since the cooling takes place as a rapid quenching with adapted cooling media such

as water, oil, polymer, or air, the brittle martensitic microstructure is formed. The higher the supercooling effect or the stronger the quenching effect, the more martensite is formed. The needle-like structures of the martensitic microstructure, as seen in Fig. 3.3, are characteristic.

For hardening, the hot work tool steel or the tool made from it is first heated to a temperature above the transformation temperature. Depending on the alloy composition of the hot work tool steel, this temperature range is approximately 850 to 1100 °C. After a defined holding time at this temperature, the "austenitization", i.e., the transformation of the initial ferritic into the austenitic microstructure, has taken place. Therefore, this transformation temperature is also called "austenitization temperature" in practice. At the same time, a large portion of the carbides, which were spherically embedded in the initial microstructure (soft annealed state), dissolve and a large part of the carbon is now dissolved in the metallic matrix. In general, the temperature for hot work tool steels is chosen so that a two-phase region of austenite and undissolved annealing carbides is present (see *Thermo-Calc diagram*, software for thermodynamic calculations of phase equilibria). This microstructure is now quenched to produce the martensitic microstructure. This occurs due to the rapid cooling, there is no time for an ordered transformation of austenite back into ferrite. The carbon dissolved in the austenite remains dissolved in the solid solution (forced dissolution), causing lattice distortions and thus a hardness in the range of approximately 55 to 65 HRC. However, this is only achieved if the rapid cooling is carried out at a cooling rate adapted to the alloy composition. In practice, this is referred to as the "critical cooling rate". This material constant can be taken from the associated time-temperature-transformation diagram (TTT diagram) and indicates the minimum cooling rate necessary for martensite formation. Based on this, the quenching medium air, oil, polymer or water (in this order with increasing quenching effect) can be selected according to ecological and economic aspects so that as little distortion and crack formation as possible occur in the hardened material. The following should apply to the choice of quenching rate: *"As fast as necessary and as slow as possible!"*.

During quenching, in addition to the martensitic transformation of the steel, carbides are again precipitated. Therefore, after quenching, the hardened structure consists of martensite and partly still residual austenite, the precipitated carbides and non-dissolved annealing carbides (Gümpel & Hoock, 1984).

Tempering

Immediately after hardening, the tempering process takes place, usually two or three times. It serves to improve the toughness and dimensional stability of hardened workpieces. These are heated again and held at different tempering temperatures for varying lengths of time. Hardening stresses are mainly reduced during this process. The brittle martensite is transformed into a structure with slightly lower hardness but somewhat higher toughness. In general, a steel becomes softer during tempering the higher it is heated (https://en.wikipedia.org/wiki/Tempering_(metallurgy)). For each steel, there are so-called tempering diagrams that show the hardness progression with increasing tempering temperature, see Fig. 3.6.

The hardness progression during tempering of some hot work tool steels is interesting. Initially, there is a slight decrease in hardness, which is due to a relaxation of the martensite lattice. At tempering temperatures above about 450 °C, finest special carbides (size of 3 to 10 nm) of the elements chromium, molybdenum, and vanadium are formed (Kulmburg, 1998). These carbides lead to precipitation hardening on the one hand and, on the other hand, the residual austenite is depleted of carbon, so the transformation temperature (i.e. the martensite finish temperature) increases. In this way, the remaining residual austenite can transform into martensite during cooling after the first tempering. This new formed martensite hardens during the second tempering due to further special carbide precipitation. The microstructure of tempered hot work tool steels thus consists of tempered martensite with non-dissolved annealing carbides and finest special carbides.

Stress-relief annealing

To ensure low-distortion further processing and to avoid the occurrence of possible hardening cracks, stress-relief annealing is therefore carried out for almost all steels, often even before hardening. Internal stresses in the semi-finished product or in the finished tool are not visible, but depending on the history of production, they are usually present. Mechanical processing (in the soft state before hardening or hard machining in the quenched and tempered state), a possibly uneven cooling after tempering, or a straightening process cause stresses in the material. Without stress-relief annealing, these stresses would be released during hardening, further processing, and finally during application, leading to geometric deviations (distortion) and possibly also to cracks. In practice, stress-relief annealing is therefore carried out during and after mechanical processing, i.e. before tempering, if necessary. Since the effect of reducing internal stresses also occurs during the multiple temperings required after hardening, additional stress-relief annealing after tempering is usually not or only rarely necessary.

In general, the heating during stress-relief annealing takes place at temperatures around 500 to 650 °C, which are always about 30 to 50 °C below the

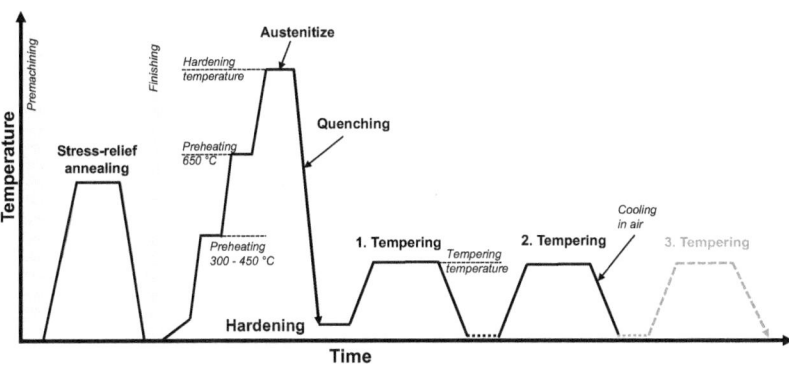

Fig. 4.5 Time-temperature sequence for stress-relief annealing, quenching and tempering of a hot work tool steel, simplified representation according to (Schruff, 2002)

tempering temperature of the particcular hot work tool steel. This avoids changes in the microstructure and thus property changes. After a holding time of usually 2 to 4 h, which is based on the size of the part to be treated, very slow cooling takes place in the furnace.

The sequence of the described heat treatment processes stress-relief annealing, quenching and tempering is shown schematically in Fig. 4.5 as a representation of the time-temperature sequence for an example in which stress-relief annealing is carried out before hardening.

4.5   Surface Treatment

A surface treatment on the finished tool always involves structuring or modification to specifically bring about property changes. This measure is carried out to increase wear resistance when the hot work tool steel alone is not sufficient for the use of the respective tool. The following are used for this purpose:

- *Changing the properties of the material through a thermochemical treatment (e.g. nitriding)*
- *Applying a wear-reducing layer to the tool surface (deposition welding)*

In addition to surface treatment processes aimed at increasing the wear resistance of new tools, worn tools are reconditioned. This reconditioning is done by welding on new material, reworking the shape contours, and repairing cracks by welding.

In all surface treatments on the finished tools, it is important to ensure that the necessary process temperatures are not higher than the previously selected tempering temperature for the hot work tool steel used. This is the only way to avoid a loss of hardness or strength of the base material.

Nitriding

Nitriding is a special process for surface hardening of steel. More precisely it is actually an enrichment of nitrogen in the tool surface through a thermochemical treatment at about 500 to 590 °C in nitrogen-containing gases (gas or plasma nitriding) or in salt baths with treatment times ranging from one hour to 100 h. In this process, nitrogen diffuses into the component surface, forming extremely hard and wear-resistant, nitride-containing layers that can be 0.2 to 0.5 mm thick, depending on the treatment. The core area of the tool being treated remains sufficiently tough and unchanged. The advantages of nitriding are that no structural transformations occur during this treatment, the resulting hard surface offers higher wear resistance, and the tendency of the hot work tool to stick and weld with the forming material is reduced. Prior to nitriding treatment, any residual stresses in the tool should be relieved by stress-relief annealing (N.N., 2018).

Deposition welding

In deposition welding, a thermal fabrication process, wear-resistant layers are applied to the workpiece surface and metallurgically bonded to the base material. In practice, this is occasionally referred to as "hard facing". Depending on the intended use of the hot work tools, conventional arc welding, laser or plasma powder arc welding (cladding) are used for coating. These methods are also used for repair welding of surfaces and edges/radii of the tools.

Applications

5

The described *hot work tool steel* is used for the chipless forming of metals at surface temperatures of the tool above 200 °C. The workpiece temperatures can range between 400 and 1200 °C. It withstands the mechanical, wear and thermal stresses that occur during tool use. Based on this, a variety of tools for hot forming and die casting are made from hot work tool steel:

- *Forging:*
  Forging saddles for open-die forging, hammer and press dies, tools for forging machines, mandrels for die forging, deburring dies, tools for press hardening
- *Extrusion:*
  Extrusion dies, die holders, extrusion punches, extrusion discs, mandrels, inner and intermediate sleeves, recipient jackets
- *Flow forming:*
  Flow forming dies, press punches, mandrels
- *Rolling:*
  Blooming-, profile-, upsetting-, bending-rolls, roll rings, pilger mandrels, rollers, mandrel bars
- *Hot cutting:*
  Hot shear blades, hot cutting plates, hot stamping tools, hot punches
- *Die casting:*
  Pressure chambers, pistons, die casting molds, ejector pins
- *Injection molding:*
  Mold tools made of plastic mold steels – see separate essential
- *Other applications:*

© The Author(s), under exclusive license to Springer Fachmedien Wiesbaden GmbH, part of Springer Nature 2024
J. Schlegel and T. Schneiders, *Hot Work Tool Steel*,
https://doi.org/10.1007/978-3-658-43016-0_5

Tools for glass production, metal powder processing such as sinter press tools, machine components for gas turbines, environmental technology, measuring devices, fittings, parts for diesel fuel pumps, pneumatic hammers and many more.

Figure 5.1 provides an overview of the application areas of various hot work tool steels, based on the hot work tool steels specified in Fig. 2.1.

For an optical impression of these diverse applications of hot work tool steels, the Fig. 5.2 shows a mosaic of selected examples.

Within the scope of this brochure only a rough classification of hot work tool steels according to their property spectrum and the requirements of their main application areas can be made from the impressive variety of applications.

Open-die forging

For open-die forging with forging hammers and forging presses, the interchangeable forging saddles are used as hot work tools. These are flat, V-shaped, or oval in shape as upper and lower saddles (see Fig. 5.2 top left). The movable upper saddle is also referred to as a "ram" in practice.

The forging saddles are subject to high heat, strong pressure, and impact stresses during use. The hot work tool steels 1.2714 (55NiCrMoV7) or 1.2779 (X6NiCrTi26-15) are predominantly used as the supporting base material for the forging saddles. Interchangeable inserts for the main wear zones, which are made for example from the highly stress able nickel-base alloy 2.4668 (NiCr19Fe19Nb5Mo3), allow for longer service life and quick tool changes when worn.

Die forging

The forging dies with the negative contours of the forging parts to be formed are also subject to high thermal loads (cyclic temperature change stress) with simultaneous strong pressure and impact stress. The service life of the forging dies depends on the complex interaction of these loads, the tool material used, the tool design, the tool guidance, the tool heat treatment and surface treatment, as well as the material to be forged. After a certain number of forging cycles, wear phenomena such as abrasion, thermal and mechanical fatigue, i.e., crack formation, and permanent deformation can occur at the contours, edges, tips, and narrow recesses (Schruff, 2002). Therefore, the well-known high requirements for the hot work tool steels used for forging dies include high thermal conductivity, high high temperature strength and hot toughness, high hardening temperature and high tempering resistance, as well as high thermal shock resistance and high

Material number	Steel short name	Hardness annealed HB max.	Hardness HRC min.	Applications
				Martensitic hot work tool steels
1.1750	C75W	217	62	Small and medium-sized dies, hot shears, riveting punches, trimming tools, profile and finishing saddles
1.2082	X21Cr13	200	(1570)*	Light alloy mold tools, pistons, pressure chambers and nozzles for light metal processing
1.2083	X40Cr14	241	52	Pressure chambers and pistons for light metal die casting with good polishability
1.2309	65MnCrMo4	(740)*	(2450)*	Blooming rolls for steel, pre-rolls for section rolling mills, upsetting rolls, bending rolls, hot rolling rings
1.2311	40CrMnMo7	230	(1770)*	Heated recipient jackets and intermediate liners in extrusion presses, die holders and inserts
1.2312	40CrMnMoS8-6	220	51	High-strength mold frames for plastics processing, tools for chipless forming
1.2313	21CrMo10	200	(1670)*	Chambers for die-casting machines, hobbed die-casting molds and similar tools
1.2323	48CrMoV6-7	220	52	Recipient jackets, intermediate boxes, die holders for extrusion presses and injection molds
1.2329	46CrSiMoV7	230	54	Matrices, molds, containers for die-casting molds, sleeves for extrusion presses (high working temperature)
1.2340	X36CrMoV5-1	200	51	Universally usable hot work tool steel for extrusion and die-casting tools
1.2342	X35CrMoV5-1-1	230	(1850)*	Mandrel bars, die-casting molds, extrusion tools
1.2343	X37CrMoV5-1	229	48	Die-casting tools, extrusion tools, plastic molds, ejector pins, forging dies
1.2344	X40CrMoV5-1	229	50	Plastic injection mold tools, extrusion tools, extrusion dies, ejectors, die-casting molds
1.2345	X50CrMoV5-1	229	60	Hot stretching rollers, scissor knives, mandrels, punches, pneumatic hammer tools and piercing tools
1.2355	50CrMoV13-15	248	56	Cold and hot forming tools, die-casting molds, powder metal dies, bending and stamping dies
1.2357	50CrMoV13-14	248	56	Cutting tools such as scissor blades, parts for diesel fuel pumps and pneumatic hammers
1.2360	X48CrMoV8-1-1	240	60	Presses, extrusion dies, die inserts with very good compressive strength
1.2362	X63CrMoV5-1	225	63	Hot-cut plates, punches, scissor knives, ejectors, trimming dies
1.2365	32CrMoV12-28	229	46	Die-casting molds, recipient inner bushings for heavy metals, press discs, pressing mandrels, perforated mandrels
1.2367	X38CrMoV5-3	229	50	High-quality dies, tools for the production of screws, nuts, rivets and bolts
1.2564	30WCrV15-1	230	52	Dies for screws, nuts, rivets, mandrels to produce hole pieces, press mandrels, press discs for non-ferrous metal processing
1.2567	30WCrV17-2	240	52	Inner bushings, press punches, press washers, mandrels, dies for heavy and light metal
1.2581	X30WCrV9-3	241	48	Recipient inner bushings, press mandrels, press dies, die-casting molds, screw and nut matrices
1.2603	45CrVMoW5-8	240	52	Hot scissor knives, inner bushings, press discs for metal extrusion presses, upsetting tools
1.2605	X35CrWMoV5	229	48	Hot work tools such as forging tools, die-casting tools, continuous casting tools, hot shears, rollers
1.2606	X37CrMoW5-1	230	58	Strand extrusion tools, pressing tools, forging dies, molding dies, die-casting molds for light metal
1.2622	X60WCrMoV9-4	270	57	Piercer plugs, mandrels for heavy metal processing
1.2662	X30WCrCoV9-3	250	52	Parts that are not cooled during hot forming: gate valves, mandrels, die-casting molds
1.2678	X45CoCrWV5-5-5	260	47	Hot extrusion dies, mandrels, punches, die and die inserts, brass die-casting molds
1.2709	X3NiCiMoTi18-9-5	323	55	Moderate thermal stress tools, die casting molds for light metal alloys
1.2711	54NiCrMoV6	240	56	Plastic molds, cutting tools, pneumatic hammer parts, diesel fuel pump parts
1.2713	55NiCrMoV6	248	54	Dies of all kinds, inserts and punches for screw production, similar tools
1.2714	55NiCrMoV7	248	42	Smaller dies, press dies, punch heads for extrusion presses, molded part press dies
1.2726	26NiCrMoV5	240	(1670)*	Pilger mandrels, plastic molds, molding plates, cold sinking, injection molding, embossing tools, bending tools
1.2738	40CrMnNiMo8-6-4	235	52	Plastic injection molds with deep engravings, e.g. for bumpers, dashboards
1.2740	28NiCrMoV10	240	49	Pilger mandrels, hot and cold cutting tools, parts for measuring instruments
1.2743	60NiCrMoV12-4	235	61	Forging dies, pressing dies of all sizes
1.2744	57NiCrMoV7-7	250	(2300)*	Dies for drop hammers, double impact hammers, press dies for light metal
1.2747	28NiMo17	258	(1860)*	Pilger mandrels
1.2766	35NiCrMo16	260	(1770)*	Molded parts dies, impact dies, hot rolling rings, inner bushings for extrusion presses, upsetting dies
1.2767	45NiCrMo16	285	52	Tools for hot pressing, cold lowering, injection molding, forming and bending, embossing tools, shearing knives, punching
1.2787	X23CrNi17	245	48	Molding tools for glass processing, pump shafts, parts for the food industry
1.2885	X32CrMoCoV3-3-3	230	54	Die casting dies, hot pressing tools, extrusion tools for heavy metals
1.2886	X15CrCoMoV10-10-5	320	50	Highly stressed hot work tools: press mandrels, press dies, hot impact extrusion tools
1.2888	X20CoCrWMo10-9	320	52	Extremely hot stressed inserts, hot pressing tools, hot extrusion tools, die casting tools
1.2889	X45CoCrMoV5-5-3	240	54	Applications with the highest demands on high temperature strength, tempering resistance and wear resistance
1.2999	X45MoCrV5-3-1	230	57	Forging dies, mandrels for high-speed forging machines, die-casting tools for heavy metals
				Austenitic hot work tool steels
1.2731	X50NiCrWV13-13			Highly stressed press dies for extrusion of heavy metals
1.2779	X6NiCrTi26-15			Inner bushings for strand extrusion of heavy metals, forging molds, die-casting molds
1.2782	X16CrNiSi25-20			Rollers for glass processing (excellent scale, corrosion resistance and high temperature strength)
1.2786	X13NiCrSi36-16			
				Nickel-base alloys
2.4668	NiCr19Fe19Nb5Mo3		(1400)*	Tools for extrusion of heavy metals, such as dies, die inserts, mandrel tips, press discs, hot scissor knives,
2.4973	NiCr19CoMo		(1300)*	sintering press tools, parts for gas turbines, environmental technology

(xxxx)* Expressed as tensile strength (N/mm^2)

Fig. 5.1 Hot work tool steels and their applications

Fig. 5.2 Mosaic of selected applications of hot work tool steels, *top left:* forging saddles of a 2000-ton press (photo: Schlegel, J., BGH Edelstahl Lippendorf GmbH), *bottom left:* hammer die for a scissor piece (photo: Beck, K.-P., Bergheim), *top right:* extrusion die (photo: VT vetimec, dies & special components), *middle right*: die casting tool (photo: VT vetimec, dies & special components), *bottom right*: low-pressure casting tool for aluminum wheels (photo: Borbet GmbH)

hot wear resistance. Based on this, the hot work tool steels 1.2344 (X40CrMoV5-1), 1.2365 (32CrMoV12-28), 1.2367 (X38CrMoV5-3), 1.2714 (55NiCrMoV7) and 1.2999 (X45MoCrV5-3-1), as well as a special molybdenum hot work tool steel according to American standard SAE-AISI H42 (T20842) are mainly used for the production of dies.

Extrusion

The extrusion process is used to produce a very long, profiled semi-finished product (extrusion) by pressing a billet (round pre-block, heated to forming temperature for steel) into a pressure chamber (recipient). The extrusion is carried out by a stamp through a shaping die, with the extrusion taking on the profile cross-section of the die. If a specially profiled mandrel is used, internally profiled hollow extrusions (tube profiles) can also be produced. Depending on the exit direction of the extrusion and the movement direction of the press stamp, a distinction is made between direct and indirect extrusion, known in practice as forward extrusion and backward extrusion. The friction forces in these process variants are very different. In forward extrusion, the internal friction between the recipient's inner wall and the surface of the block must be overcome. In contrast, this friction does not occur in backward extrusion. The friction component is even lower in the rarely used hydrostatic extrusion process. Here, the pressing force from the stamp is not applied directly to the block, but indirectly via an active medium (water or oil). The hydrostatic pressure surrounding the block in the recipient on all sides causes, at corresponding values (up to approx. 20.000 bar), the material to be pressed out through the die into an extrusion without or with very little friction in the die.

The tools for extrusion, such as the block holder (recipient) with the inner bushing, the press stamp, the press plate and die, and the mandrel, are subject to complex loads in operation, which have different effects on the individual tools, such as material fatigue, locally very high wear, and increased temperature loads at high pressures. Therefore, the hot hardness and high temperature strength with high hot toughness, very good creep resistance, and good pressure and crack resistance of the hot work tool steels used are important. For example, the following hot work tool steels are used for extrusion dies: 1.2340 (X35CrMoV5-1), 1.2343 (X37CrMoV5-1), 1.2344 (X40CrMoV5-1) and 1.2367 (X38CrMoV5-3).

Die-casting

The die-casting process involves injecting liquid metal from a casting chamber into a closed mold contour using a piston and solidifying it under pressure, usually at 200 to 300 bar. Mainly, metals with low to medium melting points such as tin, lead or zinc alloys, aluminum and magnesium, up to high-melting copper alloys are shaped using die casting. It is a very economical casting process for producing large series of formed components with high dimensional accuracy. Cold chamber and hot chamber die casting are distinguished. In the cold chamber process, the molten metal is taken from the furnace in portions and filled into the casting chamber. A hydraulically driven piston then presses this melt into the die casting mold. In this process, the cold casting chamber is only in contact with the liquid melt temporarily during casting, not continuously throughout the entire casting time. In the hot chamber die casting process, the casting chamber is constantly in contact with the melt and is therefore constantly heated to casting temperature. The requirements for the materials of the casting chamber, the casting piston, and the casting mold with its core in terms of wear resistance, tempering resistance, high temperature strength and toughness, and thermal shock crack resistance are therefore also different (https://en.wikipedia.org/wiki/Die_casting). In particular, for the cold construction of molds (basic construction with frame), the hot work tool steel 1.2312 (40CrMnMoS8-6) is used for example. Suitable hot work tool steels for the highly stressed mold-forming parts of the casting mold are: 1.2343 (X37CrMoV5-1), 1.2344 (X40CrMoV5-1), 1.2365 (32CrMoV12-28), 1.2367 (X38CrMoV5-3) or the special Cr-Mo-V alloyed steel Thermodur E 40 K Superclean from DEW. Increasingly, remelted ESR qualities are used. Often, the mold-forming components undergo surface coating after hardening to increase service life.

Material Data Sheets

6

For selected hot work tool steels, the relevant material data for each steel grade are summarized below, such as:

- *common trade names, equivalent standards and designations*
- *chemical compositions (standard analyses)*
- *physical properties*
- *information on thermal treatments, hardness progression during tempering, high-temperature strength*
- *applications*

For this selection, the most common and widely used hot work tool steel grades in practice were used. Sources included known data on these steels, which can be found in currently valid standards and material data sheets of steel manufacturers and steel dealers, in the Stahlschlüssel—Key to steel (Wegst & Wegst, 2019), as well as on Wikipedia, Wikibooks, and other encyclopedias, e.g., Metal Encyclopedia, Weltstahl.com.

Note:
Steel manufacturers often only specify one value or narrower tolerances for the contents of alloying elements in their material data sheets than the standard values of the DIN EN ISO 4957 standard allow. Such manufacturer information cannot be considered in this book, also not manufacturer-specific information, for example, for ESR-remelted high-purity grades, special steels, and other properties, such as weldability, grindability and machinability, as well as recommendations for forming, welding, favorable cutting parameters during machining and surface treatment.

© The Author(s), under exclusive license to Springer Fachmedien Wiesbaden GmbH, part of Springer Nature 2024
J. Schlegel and T. Schneiders, *Hot Work Tool Steel*,
https://doi.org/10.1007/978-3-658-43016-0_6

1.2083 (X40Cr14)

Cold work tool steel, which is also used for hot work with high hardness acceptance (hardenable steel), with high wear resistance, good machinability, good thermal conductivity, is easy to spark erode, polish and etch, shows very low distortion.

Usual steel trade names:

M310 (Böhler), **HC50** (Dörrenberg), **Formadur** 2083 (DEW)

Equivalent standads and designations:

Germany:	DIN EN ISO 4957	1.2083 (X40Cr14)	*UNS:*	
USA:	AISI / ASTM	420	*England:*	BS
Japan:	JIS		*Sweden:*	SS
France:	AFNOR	Z40C14	*Russia:*	GOST

Chemical composition (in % by mass):

	C	Si	Mn	P	S	Co	Cr	Mo	Ni	V	W	Others
min.	0,36	-	-	-	-	-	12,50	-	-	-	-	-
max.	0,42	1,00	1,00	0,030	0,030	-	14,50	-	-	-	-	-

Physical proberties:

Density ρ: 7,80 g/cm³
Specific heat capacity c: 460 J/kg·K
Modulus of elasticity E: 200 kN/mm²
Thermal conductivity λ in W/m·K: 20 °C   **21,0**
                                                        200 °C   **22,0**
                                                        300 °C   **23,8**
                                                        400 °C   **24,7**

Coefficient of thermal expansion α in 10^{-6}/K:
20 bis 100 °C   10,5
20 bis 200 °C   11,0
20 bis 300 °C   11,6
20 bis 400 °C   11,9
20 bis 500 °C
20 bis 700 °C

Thermal treatment:		*Cooling:*
Soft annealing	760 - 800 °C	≥ 3 hours, in the furnace down to 500 °C, air **Hardness annealed ≤ 230 HB**
Stress-relief annealing	600 - 650 °C	2 bis 4 hours, cooling in the furnace
Hardening	1000 - 1050 °C	Oil, compressed gas (N_2), hot bath (500 - 550 °C)
Tempering	acc. tempering diagram	

Hardness after quenching: approx. 53 HRC

Work hardness: approx. 52 HRC

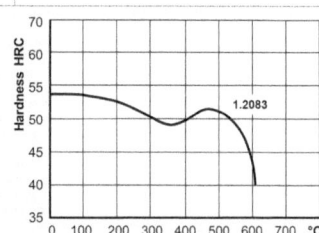

Applications:
Corrosion and acid-stressed applications as plastic mold steel, for molding and pressing tools, injection molds for abrasive plastics, machine components for the food industry, for medical technology, e.g. surgical instruments, automotive components, sensors

1.2311 (40CrMnMo7)

Quenched and tempered hot-work/plastic mold steel, delivery hardness 280 to 325 HB, low-sulfur content and pressure-resistant, with high hardenability, is easy to machine and polish, and can also be easily welded and nitrided.

Usual steel trade names:

M238 (Marks), M201 (Böhler), MCM (Dörrenberg), Formadur 2311 (DEW)

Equivalent standads and designations:

Germany:	DIN EN ISO 4957	1.2311 (40CrMnMo7)	*UNS:*	
USA:	AISI / ASTM	P20	*England:*	BS
Japan:	JIS	SKT3	*Sweden:*	SS
France:	AFNOR	40CMD8	*Russia:*	GOST

Chemical composition *(% by mass):*

	C	Si	Mn	P	S	Co	Cr	Mo	Ni	V	W	Others
min.	0,35	0,20	1,30	-	-	-	1,80	0,15	-	-	-	-
max.	0,45	0,40	1,60	0,035	0,035	-	2,10	0,25	-	-	-	-

Physical proberties:

Density ρ: 7,80 g/cm³

Specific heat capacity c: 460 J/kg·K

Modulus of elasticity E:

Thermal conductivity λ in W/m·K: 20 °C   34,5
350 °C   33,5
700 °C   32,0

Coefficient of thermal expansion α in 10^{-6}/K:

20 bis 100 °C	11,1
20 bis 200 °C	12,9
20 bis 300 °C	13,4
20 bis 400 °C	13,8
20 bis 500 °C	14,2
20 bis 600 °C	14,6
20 bis 700 °C	14,9

Thermal treatment:		*Cooling:*
Soft annealing	580 - 600 °C	≥ 3 hours, in the furnace down to 500 °C, air
		Hardness annealed ≤ 230 HB
Stress-relief annealing	approx. 600 °C	2 - 4 hours, cooling in the furnace
Hardening	830 - 870 °C	Oil, warm bath (200 - 230 °C)
Tempering	acc. tempering diagram	

Hardness after quenching: approx. 53 HRC

Work hardness: 29 - 46 HRC

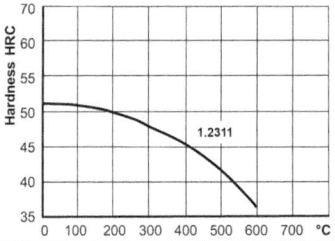

Applications:
Heated recipient jackets and intermediate sleeves in extrusion and tube presses for all metals, die holders and die inserts, parts for mechanical engineering, plastic mold making (mold plates, inserts), blow molding, injection molding and pressure molding

1.2312 (40CrMnMoS8-7)

Hot work/plastic mold steel, delivery hardness 280 to 325 HB, low alloyed with defined sulfur content, easy to machine, has good dimensional accuracy and toughness, wear-resistant after nitriding.

Usual steel trade names:

M200 (Böhler), **MCMS** (Dörrenberg), **Formadur 2312** (DEW)

Equivalent standads and designations:

Germany:	DIN EN ISO 4957	1.2312 (40CrMnMoS8-6)	*UNS:*	
USA:	AISI / ASTM	P20+S	*England:*	BS
Japan:	JIS		*Sweden:*	SS
France:	AFNOR	40CMD8.S	*Russia:*	GOST

Chemical composition *(% by mass):*

	C	Si	Mn	P	S	Co	Cr	Mo	Ni	V	W	Others
min.	0,35	0,30	1,40	-	0,050	-	1,80	0,15	-	-	-	-
max.	0,45	0,50	1,60	0,030	0,100	-	2,00	0,25	-	-	-	-

Physical proberties:

Density ρ: 7,84 g/cm^3
Specific heat capacity c: 460 J/kg·K
Modulus of elasticity E: 210 kN/mm^2
Thermal conductivity λ in W/m·K:

150 °C	**40,4**
200 °C	**40,4**
250 °C	**39,9**
300 °C	**39,0**

Coefficient of thermal expansion α in 10^{-6}/K:

20 bis 100 °C	
20 bis 200 °C	**13,0**
20 bis 300 °C	**13,7**
20 bis 400 °C	
20 bis 500 °C	
20 bis 600 °C	
20 bis 700 °C	

Thermal treatment:

Soft annealing	710 - 740 °C	4 - 6 hours, cooling in the furnace
		Hardness annealed ≤ 230 HB
Stress-relief annealing	650 - 680 °C	2 - 3 hours, cooling in the furnace
Hardening	840 - 870 °C	Oil, warm bath, air
Tempering	acc. tempering diagram	two times with two hours

Cooling: (see column above)

Hardness after quenching: approx. 54 HRC

Work hardness: approx. 34 - 50 HRC

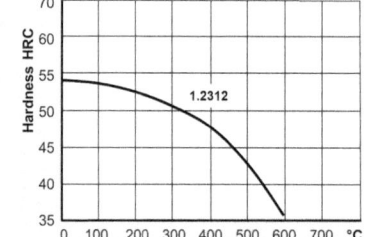

Applications:
Plastic injection molds, extruder nozzles for thermoplastics , molds and mold frames for die-casting, parts for mechanical engineering, tools for non-cutting molding

1.2329 (46CrSiMoV7)

Hot work tool steel with high tempering resistance, hot strength, high resistance to thermal shock and heat cracking, with good quenchability, with good machinability and weldability, good toughness, nitridable, PVD/CVD coated, easy to polish.

Usual steel trade names:

Thermodur 2329 (DEW)

Equivalent standads and designations:

Germany:	DIN EN ISO 4957 1.2329 (46CrSiMoV7)	
USA:	AISI / ASTM	
Japan:	JIS	
France:	AFNOR	

UNS:	
England:	BS
Sweden:	SS
Russia:	GOST

Chemical composition *(% by mass):*

	C	Si	Mn	P	S	Co	Cr	Mo	Ni	V	W	Others
min.	0,43	0,60	0,65	-	-	-	1,65	0,25	0,45	0,17	-	-
max.	0,48	0,75	0,85	0,030	0,030	-	1,85	0,35	0,60	0,22	-	-

Physical proberties:

Density ρ: 7,85 g/cm³
Specific heat capacity c:
Modulus of elasticity E:
Thermal conductivity λ in W/m·K: *150 °C*
200 °C
250 °C
300 °C

Coefficient of thermal expansion α in 10⁻⁶/K:
20 bis 100 °C
20 bis 200 °C
20 bis 300 °C
20 bis 400 °C
20 bis 500 °C
20 bis 600 °C
20 bis 700 °C

Thermal treatment:		*Cooling:*
Soft annealing	780 - 800 °C	in the furnace or in air
		Hardness annealed ≤ 230 HB
Stress-relief annealing		
Hardening	880 - 920 °C	Air, oil, warm bath (200 - 250 °C)
Tempering	acc. tempering diagram	

Hardness after quenching: approx. 53 - 55 HRC

Work hardness: approx. 50 HRC

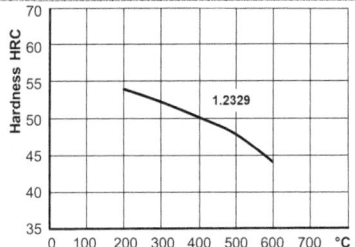

Applications:
dies, molds, stamping, engineering components for high working temperatures, low-pressure tools, containers for printing presses, sleeves for extrusion presses, extrusion press ingots

1.2340 (X36CrMoV5-1)

Hot work tool steel with very good hot strength, with improved toughness, with good thermal conductivity and insensitivity to heat cracking, high-gloss polishable, coatable, nitridable and easy to machine.

Usual steel trade names:

~ Thermodur E 38 K Superclean (DEW), W400 (Böhler)

Equivalent standads and designations:

Germany:	DIN EN ISO 4957	1.2340 (X36CrMoV5-1)	UNS:		~ T20811
USA:	AISI / ASTM	~ H11	USA:	NADCA	E1810
Japan:	JIS		England:	BS	
France:	AFNOR		Sweden:	SS	

Chemical composition *(% by mass):*

	C	Si	Mn	P	S	Co	Cr	Mo	Ni	V	W	Others
min.	0,32	-	0,10	-	-	-	4,60	1,10	-	0,35	-	-
max.	0,40	0,50	0,50	0,020	0,010	-	5,40	1,60	0,30	0,60	-	-

Physical proberties:

Density ρ: 7,80 g/cm^3
Specific heat capacity c: 460 J/kg·K
Modulus of elasticiy E: 211 kN/mm^2
Thermal conductivity λ in W/m·K:

	annealed	hardened & tempered
20 °C	29,8	26,8
350 °C	30,0	27,3
700 °C	33,4	30,3

Coefficient of thermal expansion α in 10^{-6}/K:

20 bis 100 °C	11,8
20 bis 200 °C	12,4
20 bis 300 °C	12,6
20 bis 400 °C	12,7
20 bis 500 °C	12,8
20 bis 600 °C	12,9
20 bis 700 °C	12,9

Thermal treatment:

Soft annealing	740 - 780 °C

Cooling:

in the furnace
Hardness annealed ≤ 200 HB

Stress-relief annealing	
Hardening	1000 - 1030 °C
Tempering	acc. tempering diagram

Oil, warm bath (500 - 550 °C)

Hardness after quenching: approx. 53 HRC

Work Hardness: approx. 51 - 52 HRC

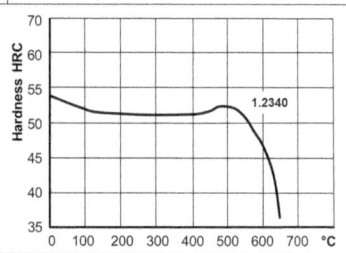

Applications:
Universally applicable hot work tool steels, especially for applications subject to high bending stresses such as extrusion and die casting tools for light metal

1.2342 (X35CrMoV5-1-1)

Hot work tool steel with high toughness, good thermal conductivity and insensitivity to heat cracking, conditionally water-coolable.

Usual steel trade names:

Thermodur 2342 EFS / 2343 EFS Superclean (DEW)

Equivalent standads and designations:

Germany:	DIN EN ISO 4957 1.2342 (X35CrMoV5-1-1)	**UNS:**	
USA:	AISI / ASTM	**USA:**	NANCA
Japan:	JIS	**England:**	BS
France:	AFNOR	**Sweden:**	SS

Chemical composition (% by mass):

	C	Si	Mn	P	S	Co	Cr	Mo	Ni	V	W	Others
min.	0,30	0,70	0,40	-	-	-	4,50	1,00	-	0,80	-	-
max.	0,40	1,20	0,60	0,030	0,030	-	5,50	1,20	-	1,00	-	-

Physical proberties:

Density ρ: 7,80 g/cm³

Specific heat capacity c:

Modulus of elasticity E:

Thermal conductivity λ in W/m·K: 20 °C   **24,5**
                                   350 °C   **26,8**
                                   700 °C   **28,8**

Coefficient of thermal expansion α in 10^{-6}/K:

20 bis 100 °C	10,9
20 bis 200 °C	11,9
20 bis 300 °C	12,3
20 bis 400 °C	12,7
20 bis 500 °C	13,0
20 bis 600 °C	13,1
20 bis 700 °C	13,5

Thermal treatment:

		Cooling:
Soft annealing	750 - 800 °C	in the furnace
		Hardness annealed ≤ 230 HB
Stress-relief annealing		
Hardening	1000 - 1040 °C	Air, oil, warm bath (500 - 550 °C)
Tempering	acc. tempering diagram	

Hardness after quenching: approx. 53 HRC

Work hardness: approx. 48 - 50 HRC

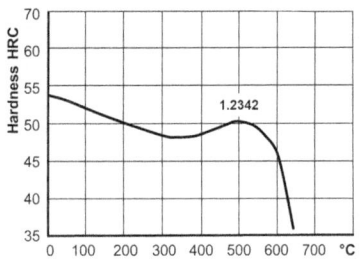

Applications:
mandrel bars, die casting molds, extrusion dies

1.2343 (X37CrMoV5-1)

Hot work tool steel with high hot strength and toughness, good thermal conductivity and insensitivity to hot cracking, conditionally water-coolable.

Usual steel trade names:

Thermodur 2343 EFS / 2343 EFS Superclean (DEW)

Equivalent standads and designations:

Germany:	DIN EN ISO 4957	1.2342 (X35CrMoV5-1-1)	UNS:	
USA:	AISI / ASTM	H11	USA:	NADCA
Japan:	JIS		England:	BS
France:	AFNOR	Z38CDV5	Sweden:	SS

Chemical composition (% by mass):

	C	Si	Mn	P	S	Co	Cr	Mo	Ni	V	W	Others
min.	0,33	0,87	0,20	-	-	-	4,80	1,10	-	0,30	-	-
max.	0,41	1,20	0,60	0,030	0,020	-	5,50	1,50	-	0,50	-	-

Physical proberties:

Density ρ: 7,80 g/cm³
Specific heat capacity c: 460 J/kg·K
Modulus of elasticity E: 215 kN/mm²
Thermal conductivity λ in W/m·K:

	annealed	hardened & tempered
20 °C	29,8	26,8
350 °C	30,0	27,3
700 °C	33,4	30,3

Wärmeausdehnungskoeffizient α in 10⁻⁶/K:

20 bis 100 °C	11,8
20 bis 200 °C	12,4
20 bis 300 °C	12,6
20 bis 400 °C	12,7
20 bis 500 °C	12,8
20 bis 600 °C	12,9
20 bis 700 °C	12,9

Thermal treatment:		Cooling:
Soft annealing	750 - 800 °C	in the furnace
		Hardness annealed ≤ 230 HB
Stress-relief annealing		
Hardening	1000 - 1030 °C	Air, oil, warm bath (500 - 550 °C)
Tempering	acc. tempering diagram	

Hardness after quenching: approx. 54 HRC

Work hardness: approx. 52 - 53 HRC

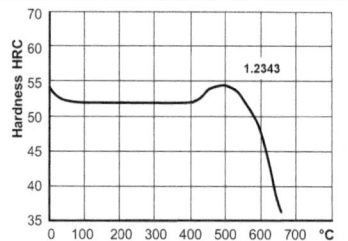

Applications:
Universally usable hot work tool steel, die-casting and extrusion tools for light metal, forging dies, mandrel bars, reinforcing rings, hot scissor blades, hot extrusion tools

1.2344 (X40CrMoV5-1)

Standard hot work tool steel with higher hot strength than 1.2343 (X37CrMoV5-1), with very good heat toughness and heat wear resistance, with good thermal conductivity and heat crack resistance.

Usual steel trade names:

Thermodur 2344 EFS / 2344 EFS Superclean (DEW), **ES 245 W** (EschmannStahl), **WP5V** (Dörrenberg)

Equivalent standads and designations:

Germany:	DIN EN ISO 4957	1.2344 (X40CrMoV5-1)	*UNS:*		T20813
USA:	AISI / ASTM	H13	*England:*	BS	BH13
Japan:	JIS	SKD61	*Sweden:*	SS	2242
France:	AFNOR	Z40CDV5	*Russia:*	GOST	4Ch5MF1S

Chemical composition *(% by mass):*

	C	Si	Mn	P	S	Co	Cr	Mo	Ni	V	W	Others
min.	0,35	0,80	0,25	-	-	-	4,80	1,20	-	0,85	-	-
max.	0,42	1,20	0,50	0,030	0,020	-	5,50	1,50	-	1,15	-	-

Physical proberties:

Density ρ: 7,78 g/cm³
Specific heat capacity c: 460 J/kg·K
Modulus of elasticity E: 215 kN/mm²
Thermal conductivity λ in W/m·K:

	annealed	hardened & tempered
20 °C	27,2	25,5
350 °C	30,5	27,6
700 °C	33,4	30,3

Coefficient of thermal expansion α in 10⁻⁶/K:

20 bis 100 °C	10,9
20 bis 200 °C	11,9
20 bis 300 °C	12,3
20 bis 400 °C	12,7
20 bis 500 °C	13,0
20 bis 600 °C	13,3
20 bis 700 °C	13,5

Thermal treatment:

		Cooling:
Soft annealing	750 - 800 °C	≥ 4 hours, in the furnace down to 500 °C, air
		Hardness annealed ≤ 240 HB
Stress-relief annealing	600 - 650 °C	in the furnace
Hardening	1000 - 1040 °C	Air, nitrogen, oil, warm bath (500 - 550 °C)
Tempering	acc. tempering diagram	

Hardness after quenching: approx. 54 HRC

Work hardness: approx. 45 - 53 HRC

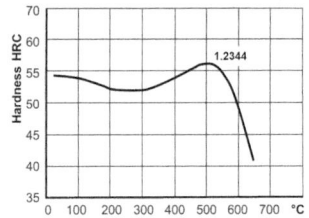

Applications:
Universally applicable, e.g. for die-casting tools and permanent molds for light metal processing, tools for forging machines, dies, die inserts, extrusion tools, mandrel rods for tube production, hot scissor knives, ejector pins

1.2345 (X50CrMoV5-1)

Hot work tool steel with increased carbon content, high wear resistance, good hot strength and high hardenability, with only little change in size.

Usual steel trade names:

DM51 (Dörrenberg), K306 (Böhler)

Equivalent standads and designations:

Germany:	DIN EN ISO 4957 1.2345 (X50CrMoV5-1)	UNS:
USA:	AISI / ASTM	England: BS
Japan:	JIS	Sweden: SS
France:	AFNOR	Russia: GOST

Chemical composition (% by mass):

	C	Si	Mn	P	S	Co	Cr	Mo	Ni	V	W	Others
min.	0,48	0,80	0,20	-	-	-	4,80	1,25	-	0,80	-	-
max.	0,53	1,10	0,40	0,030	0,030	-	5,20	1,45	-	1,00	-	-

Physical pproberties:

Density ρ: 7,80 g/cm^3
Specific heat capacity c: 460 J/kg·K
Modulus of elasticity E: 215 kN/mm^2
Thermal conductivity λ in W/m·K: 20 °C   19,5
                                                         350 °C   24,8
                                                         700 °C   26,4

Coefficient of thermal expansion α in 10^{-6}/K:

20 bis 100 °C	11,7
20 bis 200 °C	
20 bis 300 °C	12,7
20 bis 400 °C	
20 bis 500 °C	13,4
20 bis 600 °C	
20 bis 700 °C	13,8

Thermal treatment:

Soft annealing	780 - 810 °C

Cooling:

in the furnace down to 500 °C, air
Hardness annealed ≤ 230 HB

Stress-relief annealing	600 - 650 °C	2 bis 4 hours, cooling in the furnace
Hardening	1010 - 1030 °C	Air, nitrogen, oil, warm bath (500 - 550 °C)
Tempering	acc. tempering diagram	

Hardness after quenching: approx. 56 HRC

Work hardness: approx. 52 - 55 HRC

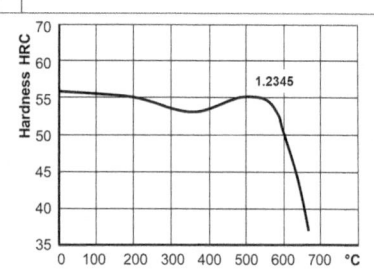

Applications:
Hot stretching rollers, scissor knives, cold pilger rollers and mandrels

1.2365 (32CrMoV12-28)

Hot work tool steel with very good hot strength, high toughness and high tempering resistance, with good thermal conductivity, high thermal shock resistance, water-coolable and cold-sinkable.

Usual steel trade names:

Thermodur 2365 EFS / 2365 EFS Superclean (DEW), **DM3** (Dörrenberg), **W320** (Böhler)

Equivalent standads and designations:

Germany:	DIN EN ISO 4957	1.2365 (32CrMoV12-28)	UNS:		T20810
USA:	AISI / ASTM	h10	England:	BS	BH10
Japan:	JIS	SKD7	Sweden:	SS	X38CrMo16
France:	AFNOR	32CDV12-28	Russia:	GOST	3Ch3M3F

Chemical composition (% by mass):

	C	Si	Mn	P	S	Co	Cr	Mo	Ni	V	W	Others
min.	0,28	0,10	0,15	-	-	-	2,70	2,50	-	0,40	-	-
max.	0,35	0,40	0,45	0,030	0,020	-	3,20	3,00	-	0,70	-	-

Physical proberties:

Density ρ: 7,85 g/cm³
Specific heat capacity c: 460 J/kg·K
Modulus of elasticity E: 215 kN/mm²
Thermal conductivity λ in W/m·K:

	annealed	hardened & tempered
20 °C	32,8	31,4
350 °C	34,5	32,0
700 °C	32,2	29,3

Coefficient of thermal expansion α in 10⁻⁶/K:

20 bis 100 °C	11,8
20 bis 200 °C	12,5
20 bis 300 °C	12,7
20 bis 400 °C	13,1
20 bis 500 °C	13,5
20 bis 600 °C	13,6
20 bis 700 °C	13,8

Thermal treatment:

Soft annealing	750 - 800 °C	in the furnace
		Hardness annealed ≤ 229 HB
Stress-relief annealing	600 - 650 °C	2 bis 4 hours, cooling in the furnace
Hardening	1030 - 1050 °C	Oil, warm bath (500 - 550 °C)
Tempering	acc. tempering diagram	

Cooling:

Hardness after quenching: approx. 52 HRC

Work hardness: approx. 50 HRC

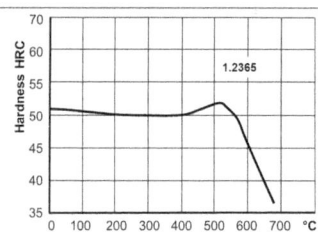

Applications:

For highly stressed hot work tools such as die-casting molds and recipient inner bushings for heavy metal alloys, press discs, press and hole mandrels for extrusion presses

1.2367 (X38CrMoV5-3)

Hot work tool steel with high hot strength and high tempering resistance, high toughness and hardenability, high thermal shock resistance and low tendency to warping

Usual steel trade names:

Thermodur 2367 EFS / 2367 EFS Superclean (DEW), **DM3X** (Dörrenberg), **W303** (Böhler)

Equivalent standads and designations:

Germany:	DIN EN ISO 4957	1.2367 (X38CrMoV5-3)	*UNS:*
USA:	AISI / ASTM		England: BS
Japan:	JIS		Sweden: SS X38CrMoV5-3
France:	AFNOR	Z38CDV5-3	Russia: GOST

Chemical composition *(% by mass):*

	C	Si	Mn	P	S	Co	Cr	Mo	Ni	V	W	Others
min.	0,35	0,30	0,30	-	-	-	4,80	2,70	-	0,40	-	-
max.	0,40	0,50	0,50	0,030	0,020	-	5,20	3,20	-	0,60	-	-

Physical properties:

Density ρ: 7,85 g/cm³

Specific heat capacity c: 460 J/kg·K

Modulus of elasticity E: 215 kN/mm²

Thermal conductivity λ in W/m·K:

	annealed	hardened % tempered
20 °C	30,8	29,8
350 °C	33,5	33,9
700 °C	35,1	35,3

Coefficient of thermal expansion α in 10^{-6}/K:

20 bis 100 °C	11,9
20 bis 200 °C	12,5
20 bis 300 °C	12,6
20 bis 400 °C	12,8
20 bis 500 °C	13,1
20 bis 600 °C	13,3
20 bis 700 °C	13,5

Thermal treatment:		*Cooling:*
Soft annealing	730 - 800 °C	in the furnace
		Hardness annealed ≤ 235 HB
Stress-relief annealing	600 - 650 °C	2 bis 4 hours, cooling in the furnace
Hardening	1020 - 1050 °C	Air, oil, warm bath (500 - 550 °C)
Tempering	acc. tempering diagram	

Hardness after quenching: approx. 57 HRC

Work hardness: 52 - 50 HRC

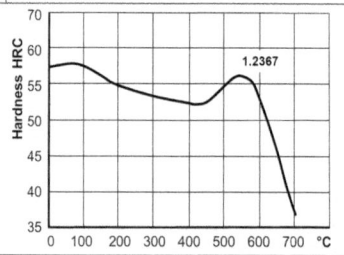

Applications:
dies, die-casting dies, intermediate bushings, die holders, press punches for heavy metal, profile dies and mandrels, tools for screw, nut, rivet and stud production, hot scissor knives

1.2606 (X37CrMoW5-1)

Hot work steel with very good hot strength and good heat-wear behavior, high thermal shock resistance.

Usual steel trade names:

1.2606 Calor MOCR (Haeckerstahl), EPS W 51 (Ossenberg)

Equivalent standads and designations:

Germany:	DIN EN ISO 4957	1.2606 (X37CrMoV5-1)	*UNS:*		
USA:	AISI / ASTM	H12	*England:*	BS	BH12
Japan:	JIS	SKD62	*Sweden:*	SS	
France:	AFNOR	Z35CWDV5	*Russia:*	GOST	

Chemical composition *(% by mass):*

	C	Si	Mn	P	S	Co	Cr	Mo	Ni	V	W	Others
min.	0,32	0,90	0,30	-	-	-	5,00	1,30	-	0,15	1,20	-
max.	0,40	1,20	0,60	0,035	0,035	-	5,60	1,60	-	0,40	1,40	-

Physical Proberties:

Density ρ: 7,85 g/cm³

Specific heat capacity c:

Modulus of elasticity E:

Thermal conductivity λ:

Coefficient of thermal expansion α in 10^{-6}/K:
20 bis 100 °C
20 bis 200 °C
20 bis 300 °C
20 bis 400 °C
20 bis 500 °C
20 bis 600 °C
20 bis 700 °C

Thermal treatment:		*Cooling:*
Soft annealing	820 - 850 °C	in the furnace
		Hardness annealed ≤ 230 HB
Stress-relief annealing		
Hardening	1000 - 1050 °C	Oil, warm bath (500 - 550 °C)
Tempering	acc. tempering diagram	

Hardness after quenching: 58 HRC

Work hardness: approx. 55 HRC

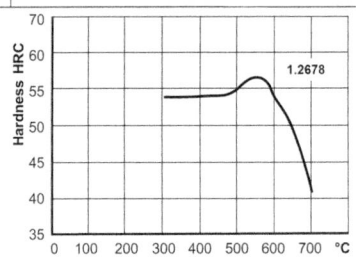

Applications:
Inner bushings and punches for metal extrusion presses, pressing tools, molding dies, forging dies, die-casting molds for light metal, hot scissor knives, trimming tools

1.2678 (X45CoCrWV5-5-5)

High-alloyed hot work tool steel with added cobalt for maximum wear resistance, with very good hot strength and tempering resistance.

Usual steel trade names:

1.2678

Equivalent standads and designations:

Germany:	DIN EN ISO 4957	1.2678 (X45CoCrWV5-5-5)	*UNS:*		T20819
USA:	AISI / ASTM	H19	*England:*	BS	
Japan:	JIS		*Sweden:*	SS	
France:	AFNOR		*Russia:*	GOST	

Chemical composition *(% by mass):*

	C	Si	Mn	P	S	Co	Cr	Mo	Ni	V	W	Others
min.	0,40	0,30	0,30	-	-	4,00	4,00	0,40	-	1,80	4,00	-
max.	0,50	0,50	0,50	0,025	0,025	5,00	5,00	0,60	-	2,10	5,00	-

Physical properties:

Density ρ: 7,70 g/cm³

Specific heat capacity c: 460 J/kg·K

Modulus of elasticity E: 200 kN/mm²

Thermal conductivity λ in W/m·K: 20 °C 25
 350 °C
 700 °C

Coefficient of thermal expansion α in 10⁻⁶/K:
 20 bis 100 °C   10
 20 bis 200 °C
 20 bis 300 °C
 20 bis 400 °C
 20 bis 500 °C
 20 bis 600 °C
 20 bis 700 °C

Thermal treatment:	*Cooling:*
Soft annealing 780 - 800 °C	in the furnace
	Hardness annealed ≤ 240 HB
Stress-relief annealing	
Hardening 1130 - 1160 °C	Oil, gas, air, warm bath,
Tempering acc. tempering diagram	

Hardness after quenching: 54 HRC

Work hardness: 50 - 54 HRC

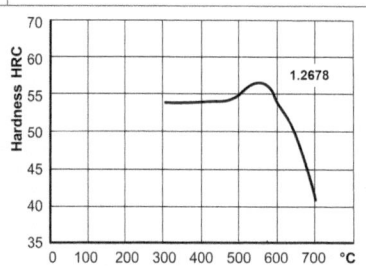

Applications:
Hot extrusion dies, mandrels, punches, highly stressed dies and die inserts, brass die-casting molds

1.2709 (X3NiCoMoTi18-9-5)

High-alloyed hot work tool steel, outstandingly low-distortion, with special property of precipitation hardenability (nickel martensite), ultra-high strength with good toughness, easy to polish, actually a cold work steel that can also be used up to 450 °C.

Usual steel trade names:

Cryodur 2709 (DEW), **18% Ni Maraging 300,** ~ W720 (Böhler)

Equivalent standads and designations:

Germany:	DIN EN ISO 4957	1.2709 (X3NiCoMoTi18-9-5)	UNS:		~ K93120
USA:	AISI / ASTM	18MAR300	England:	BS	
Japan:	JIS		Sweden:	SS	
France:	AFNOR		Russia:	GOST	

Chemical composition (% by mass):

	C	Si	Mn	P	S	Co	Cr	Mo	Ni	V	W	Others
min.	-	-	-	-	-	8,50	-	4,50	17,00	-	-	Ti
max.	0,03	0,10	0,15	0,010	0,010	10,00	0,25	5,20	19,00	-	-	0,80-1,20

Physical proberties:

Density ρ: 8,05 g/cm³

Specific heat capacity c: 460 J/kg·K

Modulus of elasticity E: 175 kN/mm²

Thermal conductivity λ in W/m·K: 20 °C   **18,4**
                                    350 °C   **23,2**
                                    500 °C   **24,0**

Coefficient of thermal expansion α in 10⁻⁶/K:

20 bis 100 °C	10,7
20 bis 200 °C	11,2
20 bis 300 °C	11,5
20 bis 400 °C	11,5
20 bis 500 °C	11,9

Thermal treatment:		*Cooling:*
Solution annealing	820 - 840 °C	Quenching in gas flow
Aging treatment	480 - 550 °C	6 hours air

Hardness: 55 - 57 HRC
(reachable by aging)

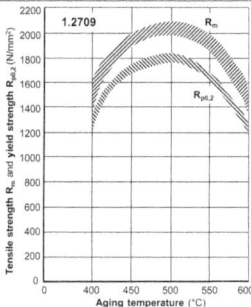

Applications:
Suitable for tools that require maximum strength and high yield and tensile strengths under moderate thermal loads, are tough and have little notch sensitivity: extrusion punches for pressing steel, diecasting molds, partial press dies, mandrels for cold rolling tubes, parts for automotive, aerospace and prototype construction, also as powder for additive applications

1.2714 (55NiCrMoV7)

Hot work tool steel, low alloyed with good toughness and high compressive strength, good hardenable in oil and air (similar: 1.2713 – 55NiCrMoV6).

Usual steel trade names:

L6-Werkzeugstahl, W500 (Böhler), Thermodur 2714 (DEW)

Equivalent standads and designations:

Germany:	DIN EN ISO 4957	1.2713 (55NiCrMoV6)	*UNS:*		T61206
USA:	AISI / ASTM	L2 / L6	*England:*	BS	
Japan:	JIS	SKT4	*Sweden:*	SS	
France:	AFNOR	55NiCrMoV7	*Russia:*	GOST	5ChNM

Chemical composition *(% by mass):*

	C	Si	Mn	P	S	Co	Cr	Mo	Ni	V	W	Others
min.	0,50	0,10	0,60	-	-	-	0,80	0,35	1,50	0,05	-	-
max.	0,60	0,40	0,90	0,030	0,030	-	1,20	0,55	1,80	0,15	-	-

Physical proberties:

Density ρ: 7,85 g/cm³
Specific heat capacity c: 470 J/kg·K
Modulus of elasticity E: 175 kN/mm²
Thermal conductivity λ in W/m·K: 20 °C   36,0
                                                    350 °C   38,0
                                                    700 °C   35,0

Coefficient of thermal expansion α in 10⁻⁶/K:
20 bis 100 °C	12,2
20 bis 200 °C	13,0
20 bis 300 °C	13,3
20 bis 400 °C	13,7
20 bis 500 °C	14,2
20 bis 600 °C	14,4

Thermal treatment:		*Cooling:*
Soft annealing	660 - 700 °C	in the furnace
		Hardness annealed ≤ 250 HB
Stress-relief annealing	630 - 650 °C	2 bis 4 hours, cooling in the furnace
Hardening	830 - 870 °C	Oil, air
Tempering	acc. tempering diagram	

Hardness after quenching: 56 - 58 HRC

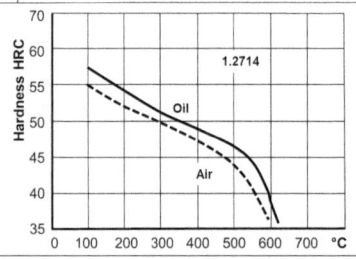

Applications:
Dies of all kinds for installation hardnesses from 355 to 410 HB as well as jaws, inserts, punches for screw production and similar tools, forging saddles, plastic press molds, rollers, rollers, hot scissor knives

1.2740 (28NiCrMoV10)

Air-hardening special steel for hot work with high toughness and thermal shock resistance.

Usual steel trade names:

Thermodur 2740 (DEW)

Equivalent standads and designations:

Germany:	DIN EN ISO 4957	1.2740 (28NiCrMoV10)	**UNS:**
USA:	AISI / ASTM		*England:* BS
Japan:	JIS		*Sweden:* SS
France:	AFNOR		*Russia:* GOST

Chemical composition *(% by mass):*

	C	Si	Mn	P	S	Co	Cr	Mo	Ni	V	W	Others
min.	0,24	0,30	0,20	-	-	-	0,60	0,50	2,30	0,25	-	-
max.	0,32	0,50	0,40	0,030	0,030	-	0,90	0,70	2,60	0,32	-	-

Physical proberties:

Density ρ: 7,85 g/cm³

Specific heat capacity c:

Modulus of elasticity E:

Thermal conductivity λ:

Coefficient of thermal expansion α in 10^{-6}/K:
20 bis 100 °C
20 bis 200 °C
20 bis 300 °C
20 bis 400 °C
20 bis 500 °C
20 bis 600 °C

Thermal treatment:		*Cooling:*
Soft annealing	670 -7800 °C	in the furnace **Hardness annealed ≤ 240 HB**
Stress-relief annealing		
Hardening	840 - 870 °C	Oil, air
Tempering	acc. tempering diagram	

Hardness after quenching: 49 HRC

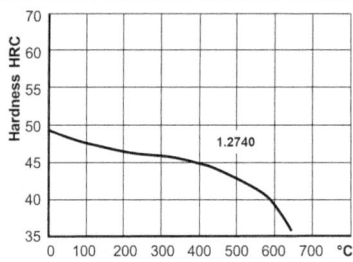

Applications:
Special steel for mandrel rods and pilgrim mandrels

1.2766 (35NiCrMo16)

Tool steel for hot work, easy to harden, low distortion, with high toughness, polishable.

Usual steel trade names:

35NiCrMo16 or: X35NiCrMo4

Equivalent standads and designations:

Germany:	DIN EN ISO 4957 1.2766 (35NiCrMo16)	**UNS:**
USA:	AISI / ASTM	England: BS
Japan:	JIS	Sweden: SS
France:	AFNOR	Russia: GOST

Chemical composition *(% by mass):*

	C	Si	Mn	P	S	Co	Cr	Mo	Ni	V	W	Others
min.	0,32	0,15	0,40	-	-	-	1,20	0,20	3,80	-	-	-
max.	0,38	0,30	0,60	0,035	0,035	-	1,50	0,40	4,30	-	-	-

Physical proberties:

Density ρ: 7,85 g/cm³
Specific heat capacity c:
Modulus of elasticity E:
Thermal conductivity λ in W/m·K: 20 °C
                                  250 °C   **42,2**
                                  850 °C   **42,3**

Coefficient of thermal expansion α in 10⁻⁶/K:
20 bis 100 °C
20 bis 200 °C
20 bis 250 °C   24
20 bis 400 °C
20 bis 500 °C
20 bis 600 °C

Thermal treatment:	*Cooling:*
Soft annealing 620 - 660 °C	in the furnace
	Hardness annealed ≤ 260 HB
Stress-relief annealing	
Hardening 820 - 850 °C	Oil, air
Tempering acc. tempering diagram	

Hardness after quenching: 56 HRC

Work hardness: 37 - 49 HRC

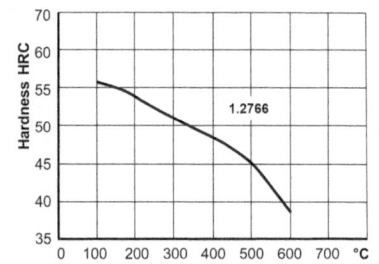

Applications:
Highly stressed press and impact dies, press punches, upsetting tools, hot rolling rings and inner bushings for metal extrusion presses

1.2767 (45NiCrMo16)

Quenched and tempered steel (good hardenability), nickel alloyed, with highest toughness, very high compressive and flexural strength, high-gloss polishable. Also used as cold work steel.

Usual steel trade names:

45NiCrMo16 or: X45NiCrMo4

Equivalent standads and designations:

Germany:	DIN EN ISO 4957	1.2767 (45NiCrMo16)	*UNS:*	
USA:	AISI / ASTM	~ 6F7	*England:*	BS
Japan:	JIS		*Sweden:*	SS
France:	AFNOR	45NCD16	*Russia:*	GOST

Chemical composition (% by mass):

	C	Si	Mn	P	S	Co	Cr	Mo	Ni	V	W	Others
min.	0,40	0,10	0,20	-	-	-	1,20	0,15	3,80	-	-	-
max.	0,50	0,40	0,50	0,030	0,030	-	1,50	0,35	4,30	-	-	-

Physical proberties:

Density ρ: 7,85 g/cm³
Specific heat capacity c: 460 J/kg·K
Modulus of elasticity E: 210 kN/mm²
Thermal conductivity λ in W/m·K: 20 °C   28
                                  1 00 °C   30

Coefficient of thermal expansion α in 10⁻⁶/K:
20 bis 100 °C
20 bis 200 °C
20 bis 250 °C
20 bis 400 °C
20 bis 500 °C
20 bis 600 °C

Thermal treatment:		*Cooling:*
Soft annealing	610 - 650 °C	in the furnace
		Hardness annealed ≤ 260 HB
Stress-relief annealing	approx. 650 °C	in the furnace
Hardening	840 - 870 °C	Oil, warm bath, air
Tempering	acc. tempering diagram	

Hardness after quenching: 53 - 57 HRC

Work hardness: 52 - 55 HRC

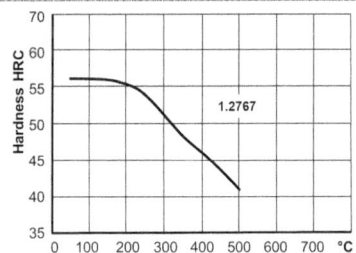

Applications:
Plastic molds, mold plates, solid stamping tools, mold inserts for injection molds, embossing, forming and bending tools, cold sinking tools

1.2782 (Xx16CrNiSi25-20)

Austenitic hot work tool steel, scale resistant to air up to 1150 °C, resistant to oxidizing atmosphere, excellent hot strength, good cold forming.

Usual steel trade names:

Thermodur 2782 (DEW), TK 2782 ESU (thyssenkrupp)

Equivalent standads and designations:

Germany:	DIN EN ISO 4957 1.2782 (X16CrNiSi25-20)	UNS:	
USA:	AISI / ASTM	England:	BS
Japan:	JIS	Sweden:	SS
France:	AFNOR	Russia:	GOST

Chemical composition (% by mass):

	C	Si	Mn	P	S	Co	Cr	Mo	Ni	V	W	Others
min.	-	1,80	-	-	-	-	24,00	-	19,00	-	-	-
max.	0,20	2,30	2,00	0,035	0,035	-	26,00	-	21,00	-	-	-

Physical proberties:

Density ρ: 7,85 g/cm³
Specific heat capacity c: J/kg·K

Modulus of elasticity E: 190 - 210 kN/mm²

Thermal conductivity λ in W/m·K: 20 °C   13,0
                                    500 °C   19,0

Coefficient of trhermal expansion α in 10^{-6}/K:
20 bis 100 °C
20 bis 200 °C   16,5
20 bis 250 °C
20 bis 400 °C   17,0
20 bis 500 °C
20 bis 600 °C   17,5

Thermal treatment:	Cooling:
Solution annealing 1000 - 1100 °C	Air or water
Aging treatment	Hardness annealed ≤ 230 HB

Strength after quenching: 495 - 705 MPa

Applications:
Tools for glass processing, e.g. caps, pipe bowls, pipe spindles, mouthpieces, starting iron tools

1.2787 (X23CrNi17)

Hot work tool steel, temperable, resistant to corrosion and scale.

Usual steel trade names:

Thermodur 2787 (DEW), N350 (Böhler)

Equivalent standads and designations:

Germany:	DIN EN ISO 4957	1.2787 (X23CrNi17)	UNS:		
USA:	AISI / ASTM		England:	BS	S80
Japan:	JIS	SUS431FB / SUS431	Sweden:	SS	
France:	AFNOR	Z15CN16-02	Russia:	GOST	~ 20Ch17N2

Chemical composition (% by mass):

	C	Si	Mn	P	S	Co	Cr	Mo	Ni	V	W	Others
min.	0,10	-	-	-	-	-	15,50	-	1,00	-	-	-
max.	0,25	1,00	1,00	0,035	0,035	-	18,00	-	2,50	-	-	-

Physical proberties:

Density ρ: 7,70 g/cm³
Specific heat capacity c: 460 J/kg·K
Modulus of elasticity E: 215 kN/mm²
Thermal conductivity λ in W/m·K: 20 °C   25

Coefficient of thermal expansion α in 10⁻⁶/K:
20 bis 100 °C	10,0
20 bis 200 °C	10,5
20 bis 300 °C	11,0
20 bis 400 °C	11,0
20 bis 500 °C	11,0

Thermal treatment:

Cooling:

Soft annealing	710 - 750 °C	in the furnace
		Hardness annealed ≤ 245 HB
Stress-relief annealing	approx. 650 °C	in the furnace
Hardening	990 - 1020 °C	Oil or warm bath, 200 °C
Tempering	acc. tempering diagram	

Hardness after quenching: 47 HRC

Work hardness: 38 - 45 HRC

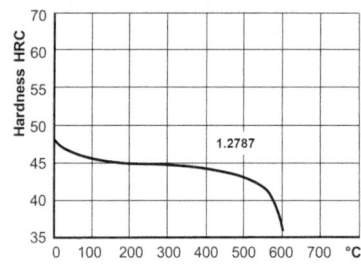

Applications:
Tools for glass processing

1.2885 (X32CrMoCoV3-3-3)

Hot work tool steel with cobalt content, with good hot strength, good tempering resistance and good heat-wear resistance, with good thermal conductivity it can tolerate hard water cooling.

Usual steel trade names:

LO-W 2885 (Lohmann)

Equivalent standads and designations:

Germany:	DIN EN ISO 4957	1.2885 (X23CrMoCoV3-3-3)	*UNS:*	
USA:	AISI / ASTM	H10A	*England:*	BS
Japan:	JIS		*Sweden:*	SS
France:	AFNOR		*Russia:*	GOST

Chemical composition *(% by mass):*

	C	Si	Mn	P	S	Co	Cr	Mo	Ni	V	W	Others
min.	0,28	0,10	0,15	-	-	2,50	2,70	2,60	-	0,40	-	-
max.	0,35	0,40	0,45	0,030	0,030	3,00	3,20	3,00	-	0,70	-	-

Physical proberties:

Density ρ: 7,88 g/cm³

Specific heat capacity c:

Modulus of elasticity E:

Thermal conductivity λ in W/m·K: *20 °C*   **27,3**

Coefficient of thermal expansion α in 10^{-6}/K:

20 bis 100 °C	10,5
20 bis 200 °C	11,3
20 bis 300 °C	11,8
20 bis 400 °C	12,3
20 bis 500 °C	12,5
20 bis 600 °C	12,8

Thermal treatment:	*Cooling:*
Soft annealing 760 - 840 °C	in the furnace down to < 500 °C
	Hardness annealed ≤ 230 HB

Stress-relief annealing	
Hardening 1000 - 1050 °C	Oil, gas, warm bath (550 °C)
Tempering acc. tempering diagram	

Hardness after quenching: 52 HRC

Work hardness: 46 - 50 HRC

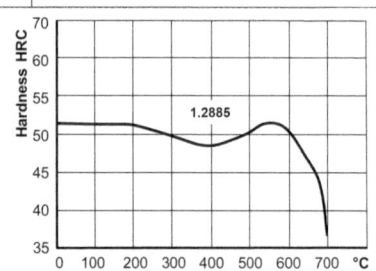

Applications:
Tools for die-casting, hot pressing, extrusion and continuous casting mainly for heavy metals, hole mandrels

1.2888 (X20CoCrWMo10-9)

High-alloyed hot work tool steel with special resistance to high temperature wear, has extremely high tempering resistance and high temperature resistance to molten metals.

Usual steel trade names:

LO-W 2888 (Lohmann)

Equivalent standads and designations:

Germany:	DIN EN ISO 4957	1.2888 (X20CoCrWMo10-9)	*UNS:*	
USA:	AISI / ASTM		*England:*	BS
Japan:	JIS		*Sweden:*	SS
France:	AFNOR		*Russia:*	GOST

Chemical composition *(% by mass):*

	C	Si	Mn	P	S	Co	Cr	Mo	Ni	V	W	Others
min.	0,17	0,15	0,40	-	-	9,50	9,00	1,80	-	-	5,00	-
max.	0,23	0,35	0,60	0,035	0,035	10,50	10,00	2,20	-	-	6,00	-

Physical properties:

Densitiy ρ: 8,08 g/cm³

Specific heat capacity c:

Modulus of elasticity E: 215 kN/mm²

Thermal conductivity λ in W/m·K: *20 °C* **15,9**

Wärmeausdehnungskoeffizient α in 10⁻⁶/K:
- *20 bis 100 °C*
- *20 bis 200 °C*
- *20 bis 300 °C*
- *20 bis 400 °C*
- *20 bis 500 °C*
- *20 bis 600 °C*

Thermal treatment:		*Cooling:*
Soft annealing	760 - 880 °C	in the furnace down to < 500 °C **Hardness annealed ≤ 340 HB**
Stress-relief annealing	600 - 650 °C	cooling in air
Hardening	1100 - 11650 °C	Oil, gas, warm bath (550 °C)
Tempering	acc. tempering diagram	

Hardness after quenching: 52 HRC

Work hardness: 42 - 54 HRC

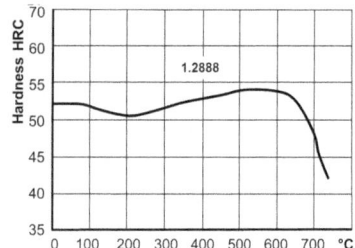

Applications:

Extrusion tools for copper, copper alloys and steels, dies, die-casting tools of all kinds for brass, chambers for magnesium die-casting

Literature

Bauer, G. et al. (2000). *Vanadium and Vanadium Compounds*. In: Ullmann's Encyclopedia of Industrial Chemistry. Wiley-VCH, Weinheim.

Bayer, E. & H. Seilstorfer (1984). *Pulvermetallurgisch durch Heißisostatisches Pressen hergestellter Warmarbeitsstahl X40CrMoV5-1*. Archiv für das Eisenhüttenwesen 55 / 4, pp. 169–176.

Becker, H.-J. & F. Kiel (1983). *Schadensfälle bei Werkzeugen, Ermittlung der Ursachen und Hinweise zu ihrer Vermeidung*. Thyssen Stainless Steel Technical Reports 9 / 2, pp. 171–187.

Berns, H. (1993). *Stahlkunde für Ingenieure*. Springer Verlag Berlin Heidelberg New York, 2nd Edition.

Berns, H. (2004). *Beispiele zur Schädigung von Warmarbeitswerkzeugen*. Heat Treatment Technical Communications 59 / 6, pp. 379–387.

Bockholt, D. (2002). *Charakterisierung des Eigenschaftsprofils eines neuen pulvermetallurgisch hergestellten Warmarbeitsstahles im Vergleich zu herkömmlichen Standardwarmarbeitsstählen*. Diploma Thesis, Institute for Materials, Ruhr University Bochum.

Burghardt, H. & G. Neuhof (1982). *Stahlerzeugung*. VEB Deutscher Verlag für Grundstoffindustrie, Leipzig.

Ehrhardt, R. (2008). *Warmarbeitsstähle – Hochwertige Edelstähle von Deutsche Edelstahlwerke*. DEW-Sales Trainings, 24.09.2008.

Ernst, C. (2009). *150 Jahre Werkzeugstahl: ein Werkstoff mit Zukunft. Prozess- und legierungstechnische Entwicklung bei der (Werkzeug) Stahlerzeugung*. Zeitschrift Ferrum: Nachrichten aus der Eisenbibliothek, Stiftung der Georg Fischer AG, 81, pp. 66-76

Gottstein, G. (2014). *Materialwissenschaft und Werkstofftechnik - Physikalische Grundlagen*. 4th Edition. Springer Vieweg Berlin, Heidelberg.

Grinder, O. (1999). *PM HSS and Tool Steels – Present State of the Art and Development Trends*. In Jeglitsch, F.; Ebner, R. & H. Leitner: Proceedings of the 5th International Tooling Conference: *Tool Steels in the Next Century*, Leoben, Austria, pp. 39–47.

Gümpel, P. (1983). *Untersuchungen über Primärkarbide in Warmarbeitsstählen*. Thyssen Edelstahl Technische Berichte 9 / 2, pp. 121–123.

Gümpel, P. & M. Hoock. (1984). *Carbidausscheidungen in Warmarbeitsstählen*. Archiv für das Eisenhüttenwesen 55 / 10, pp 493–498.

Huemer, K., Wolf, G., Sormann, A. et al. (2005). *Auswirkungen einer Kalziumbehandlung auf die Entstehung und Zusammensetzung von nichtmetallischen Einschlüssen bei*

J. Schlegel and T. Schneiders, *Hot Work Tool Steel*,
https://doi.org/10.1007/978-3-658-43016-0

der Erzeugung von aluminiumberuhigten Stählen für Langprodukte. BHM Berg- und Hüttenmännische Monatshefte 150, pp. 237-242.

IHT (2022). *Tiefkühlen.* Technische Information - Industrieverband Härtetechnik e. V.

Issler, L., H. Ruoß & P. Häfele (2003). *Festigkeitslehre - Grundlagen.* Springer-Verlag Berlin Heidelberg.

Johannsen, O. (1953). *Geschichte des Eisens.* 3rd Edition, Verlag Stahleisen.

Jung, I. (2003). *Neue Hochleistungsstähle – Neue Trends, Herstellverfahren, Eigenschaften und Anwendungen für den Werkzeugbau.* Stahl 5 / 6, pp. 41–43.

Karagöz, S. & H.-O. Andrén (1992). *Secondary Hardening in High-Speed Steels.* Journal of Metallurgy 83 / 6, pp. 386–394.

König, F. & W. Klocke (2006). *Fertigungsverfahren 4 - Umformen,* Springer, 5th Edition.

Kulmburg, A. (1998). *Das Gefüge der Werkzeugstähle – ein Überblick für den Praktiker. Teil 1: Einteilung, Systematik und Wärmebehandlung der Werkzeugstähle.* Praktische Metallographie 35 / 4, pp. 180–202.

Kulmburg, A.; Schindler, A.; Fauland, H.P. & G. Hackl (1994): *Der Einfluß der Herstellbedingungen auf die Zähigkeit von Werkzeugstählen.* Härterei Technische Mitteilungen 49 / 1, pp. 31–39.

Langehenke, H. (2007). *Werkstoff-Kurznamen und Werkstoff-Nummern für Eisenwerkstoffe: DIN-Normenheft 3 DIN-Normen und Werkstoffblätter Querverweislisten,* Taschenbuch, Beuth.

Liedtke, D. (2005). *Wärmebehandlung von Stahl – Härten, Anlassen, Vergüten, Bainitisieren,* Wirtschaftsvereinigung Stahl, Merkblatt 450 (Edition 2005).

Macherauch, E. & HW. Zoch (2011). *Reibung und Verschleiß.* In: Praktikum in Werkstoffkunde. Vieweg+Teubner.

Meyer, W., Hochörtler, J. & A. Kucharz (1995). *Entwicklung auf dem Gebiet der Schmelz- und Sekundärmetallurgie zur Eigenschaftsverbesserung spezieller Stahlqualitäten.* BHM Berg- und Hüttenmännische Monatshefte 140, pp. 4-14.

N.N. (1994). *Modern methods for the manufacture of tool steels.* Steel Times (1994), pp. 359-360.

N.N. (2018). *Warmarbeitsstahl.* Technical Information Voestalpine Böhler Edelstahl GmbH & Co KG (BW015DE – 05.2018).

Persson, A. et al. (2002). *Influence of Surface Engineering on the Performance of Tool Steels for Die Casting.* In Bergström, J.; Frederiksson, G.; Johansson, M.; Kotik, O. & F. Thuvander: *The Use of Tool Steels: Experience and Research.* Proceedings of the 6th International Tooling Conference, Karlstad, Sweden, Volume 2, pp. 841–854.

Schlegel, J. (2023). *The World of Steel.* Springer. (ISBN 978-3-658-39732-6)

Schneiders, Till (2006). *Neue pulvermetallurgische Werkzeugstähle.* Fortschritt-Berichte VDI Reihe 5 Nr. 721, VDI-Verlag, Düsseldorf, PhD thesis, Institute for Materials, Faculty of Mechanical Engineering, Ruhr University Bochum.

Schruff, I. (2002). *Zusatzstudium Stahl: Technologie der Werkzeugstähle,* Edelstahl Witten-Krefeld GmbH (IS 003 2002).

Schruff, I. (1989). *Zusammenstellung der Eigenschaften und Werkstoffkenngrößen der Warmarbeitsstähle X38CrMoV5-1 (Thyrotherm 2343), X40CrMoV5-1 (Thyrotherm 2344), X32CrMoV3-3 (Thyrotherm 2365) und X38CrMoV5-3 (Thyrotherm 2367).* Thyssen Edelstahl Technische Berichte 15 / 2, pp. 70–81.

Schruff, I. et all. (2003). *Formen und Werkzeuge für hohe Standzeiten*. stahl und eisen 123 / 4, pp. 75–80.

Skolaut, W. (Hrsg.). (2018). *Maschinenbau*. technical book, 2nd Updated Edition. Springer Vieweg.

Spur, G. (1991). *Vom Wandel der industriellen Welt durch Werkzeugmaschinen*. Carl Hanser Verlag. München-Wien.

Trenkler, H. & W. Kreiger (1988). *Gmelin–Durrer: Metallurgy of Iron*. Volume 9 (Practice of Steelmaking). 4th Edition, Springer Verlag Berlin.

Trent, E. M., & P. K. Wright (2000). *Metal cutting*. Butterworth-Heinemann, Boston, 4th Edition.

Wegst, C., & M. Wegst (2019). *Stahlschlüssel – key to steel*. Stahlschlüssel Wegst GmbH.

Weißbach, W. (2012). *Werkstoffkunde: Strukturen, Eigenschaften, Prüfung*. 18th Edition. Vieweg + Teubner, Wiesbaden.

Wendl, F. (1985). *Einfluss der Fertigung auf Gefüge und Zähigkeit von Warmarbeitsstählen mit 5 % Chrom*, Fortschritt-Berichte VDI Reihe 5 Nr. 91, VDI-Verlag, Düsseldorf. PhD thesis, Institute for Materials, Faculty of Mechanical Engineering, Ruhr University Bochum.

Weißbach, W. (2007). *Werkstoffkunde: Strukturen, Eigenschaften, Prüfung*. 16th Updated Edition. Friedr. Vieweg & Sohn Verlag GWV Fachverlage GmbH, Wiesbaden.

Wilmes, S. (1990). *Pulvermetallurgische Werkzeugstähle – Herstellung, Eigenschaften und Anwendung*. Stahl und Eisen 1.

https://en.wikipedia.org/wiki/Tempering_(metallurgy) (Accessed: 04.06.2022)

https://en.wikipedia.org/wiki/Annealing_(materials_science) (Accessed: 03.06.2022)

https://en.wikipedia.org/wiki/Hardening_(metallurgy) (Accessed: 03.06.2022)

https://en.wikipedia.org/wiki/Iron_Age (Accessed: 01.06.2022)

https://en.wikipedia.org/wiki/Die_casting (Accessed: 09.06.2022)